Introduction to Workplace Safety and Health Management

A Systems Thinking Approach

About the Author

Goh Yang-Miang is Associate Professor and Deputy Head (Research) at the Department of Building, School of Design and Environment, National University of Singapore (NUS). He holds both a BEng and PhD in Civil Engineering from NUS. Prior to his current appointment, Dr Goh was Deputy Programme Director (B.Sc. Project and Facilities Management), Department of Building, NUS (2013–2016); Assistant Professor at the Department of Building, NUS (2012–2016); Senior Consultant at Det Norske Veritas, Singapore (2011–2012); Senior Lecturer at the School of Public Health, Curtin University, Western Australia (2007–2010); Engineer, Group Leader, Head, and Assistant Director (Investigations) at the Occupational Safety and Health Division at the Ministry of Manpower, Singapore (2005–2007); and Research Engineer/Research Assistant at NUS (2003–2004).

Prof Goh is also a Member of the External Review Panel on SAF Safety (ERPSS) (2017–present); Member of the Workplace Safety and Health Council Construction and Landscape Committee (2015–2017); Chairman/Co-Chairman of the Health and Safety Engineering Technical Committee, Institution of Engineers (IES), Singapore (2015–2018); and Academic Panel Member, IES Academy (2016–present). His current research interests are: Workplace safety and health; Systems thinking; Fall protection; Analytics, machine learning and knowledge-based systems; and Project and construction management.

Introduction to Workplace Safety and Health Management

A Systems Thinking Approach

Goh Yang Miang

W World Scientific

NEW JERSEY · LONDON · SINGAPORE · BEIJING · SHANGHAI · HONG KONG · TAIPEI · CHENNAI

Published by

World Scientific Publishing Co. Pte. Ltd.

5 Toh Tuck Link, Singapore 596224

USA office: 27 Warren Street, Suite 401-402, Hackensack, NJ 07601

UK office: 57 Shelton Street, Covent Garden, London WC2H 9HE

National Library Board, Singapore Cataloguing in Publication Data
Names: Goh, Yang Miang.
Title: Introduction to workplace safety and health management : a systems thinking approach /
    Goh Yang Miang.
Description: Singapore : World Scientific, [2018] | Includes bibliographic references.
Identifiers: OCN 1045423262 | 978-981-32-7425-9 (paperback) | 978-981-32-7411-2 (hardcover)
Subjects: LCSH: Industrial safety--Management. | Industrial hygiene--Management.
Classification: DDC 658.382--dc23

British Library Cataloguing-in-Publication Data
A catalogue record for this book is available from the British Library.

For any available supplementary material, please visit
https://www.worldscientific.com/worldscibooks/10.1142/11094#t=suppl

Desk Editor: Amanda Yun

Typeset by Stallion Press
Email: enquiries@stallionpress.com

Contents

1

Introduction

1.1 Importance of Workplace Safety and Health Management

Neglecting workplace safety and health[1] (WSH) management can cause accidents and ill health, which can lead to severe consequences for individuals, families, communities and organisations. However, as WSH incidents are uncertain events which may not occur even if work is conducted unsafely, many organisations do not see the importance of WSH. Managers may also neglect WSH management when there are other pressing issues related to revenue, schedule and client expectations that require their attention and resources.

This book is written for current and future managers overseeing and managing high risk workplaces like construction projects, shipyards and factories. Even if a manager is not managing operations directly, s/he makes decisions that have significant influence on WSH. Thus, this book is also relevant to professionals like project managers working for clients, contracts managers, engineers and architects, who are involved in the design, selection and planning of products and operations. Knowledge of WSH management will also allow managers to identify contractors with a strong WSH competency. In addition, with legislations such as the WSH (Design for Safety) Regulations ('DfS Regulations'), all stakeholders are expected to be proactive in WSH management. Design for Safety (DfS) (also known as prevention through design, safe design and Construction (Design and Management)) promotes early consideration of safety and health hazards during the design phase of a construction

---

[1] Also known as occupational safety and health (OSH) or occupational health and safety (OHS).

project. Similarly, in other industries, considerations of safety during upstream stages have significant benefits. With early intervention, hazards can be more effectively eliminated or controlled leading to safer workplaces and processes.

1.1.1 Accidents

Accidents happen, and when they happen, people suffer. However, many people simply do not register this simple fact. Even in relatively low risk environments like universities, there have been incidents of fire, which have caused property damage. For example, the National University of Singapore (NUS) had fires on in August 2012, October 2012 and April 2014,[2] fortunately, these fires did not result in any major injury or fatality. In 2008, a tower crane collapsed in NUS,[3] killing three workers. In Singapore, the Nicoll Highway collapsed on 20 April 2004 (Goh and Soon 2014), and it remains one of the worst industrial accidents that Singapore has had. The Nicoll Highway collapse led to four deaths, and the body of the foreman, Mr Heng Yeow Peow, who saved several workers during the collapse, was never found. Though the four deaths were very significant, and the impact on the families of the victims was immeasurable, it must be remembered that the scale of the collapse of the Nicoll Highway, a busy road linking to the central area of Singapore, could have easily led to more fatalities. After the accident, senior managers of the main contractor, Nishimatsu (partner in the Nishimatsu-Lum Chang Joint Venture), were severely fined (Goh 2008). The Professional Engineer (PE) cum project coordinator had made severe errors in the design of the diaphragm wall and strut system, which led to the collapse of the cut and

---

[2] News articles on the fires can be seen online at https://sg.news.yahoo.com/fire-breaks-out-at-nus.html, https://sg.news.yahoo.com/small-fire-breaks-out-in-nus-lecture-theatre.html, and http://www.asiaone.com/singapore/fire-nus-engineering-faculty-building.

[3] News article available on http://www.asiaone.com/News/Education/Story/A1Story20080222-50958.html.

cover tunnel. The PE from New Zealand was fined S\$160,000 in April 2006 and banned from practice for two years. In addition, Nishimatsu was fined S\$200,000, the design manager was fined S\$160,000 and the project director was fined S\$120,000. The former Land Transport Authority (LTA) project director was fined S\$8,000 for failing to exercise due diligence in monitoring excavation works and assessing readings of soil monitoring instruments. The Nicoll Highway station project was delayed by four years and millions of dollars were spent on reconstructing the highway and adjacent roads. The Nicoll Highway collapse showed not only that accidents cause human suffering and project failure, but that project managers and executives can be taken to task for not ensuring safety.

In recent years, the Singapore Ministry of Manpower (MOM) has been taking project managers and appointment holders to task for failure to ensure safety on site. In one instance, a project coordinator cum lifting supervisor was sentenced to four weeks' imprisonment. Thus, managers must be competent in and committed to WSH management.

1.1.2 Ill health

In contrast to safety, occupational health is frequently neglected in many industries. Unlike accidents, ill health takes years to develop and the illness may not be clearly linked to the work that the victim conducted many years ago. According to the International Labor Organization (2009), each year, about 2.3 million people die from work-related accidents and diseases. Out of the 2.3 million deaths, about 360,000 are due to accidents, while the remaining 1.95 million are fatal work-related diseases. Even though most of the deaths occur in developing countries and the statistics include all types of workplace, the astonishing numbers signal the impact of occupational diseases. Some of the occupational health issues common in the industries include excessive noise, skin dermatitis, musculoskeletal disorders, cancer, and occupational asthma.

1.1.3 Environmental pollution

Environmental pollution includes local pollution problems like noise and odour, and global pollution issues like the depletion of natural resources, inefficient use of energy, emission of carbon dioxide and greenhouse gases, air pollution, water pollution, land pollution, and loss of biodiversity. Both local and global environmental pollution are undesirable, but global pollution, particularly with regard to climate change-related issues, is becoming more and more important. With the global average temperature on the rise and more extreme and erratic weather occurring, the industry has to study the environmental aspects (synonymous to hazards in WSH terminology) and impacts (synonymous to accidents or ill health) carefully.

Environmental problems are closely related to WSH. Accidents such as major fires and structural collapses can release substances and dust into the environment and will also lead to pollution and other environmental impacts. As in the case for WSH, the Singapore National Environmental Agency (NEA) has been very strict in their enforcement of Environmental Protection and Management legislations and construction sites have to ensure they have a systematic approach for environmental management.

1.1.4 Cost of incidents

WSH incidents are costly. According to Bird *et al.* (2003) (see Figure 1.1), for each dollar of insured cost, there is another $6 to $53 of uninsured costs. Just like a ship captain who only sees the tip of the iceberg, many organisations are focused on the insured costs, but they may not have noticed the massive hidden uninsured costs.

1.2 Workplace Safety and Health Statistics

WSH statistics help us understand how different industries are performing in terms of WSH. The WSH statistics in Singapore are reported by the

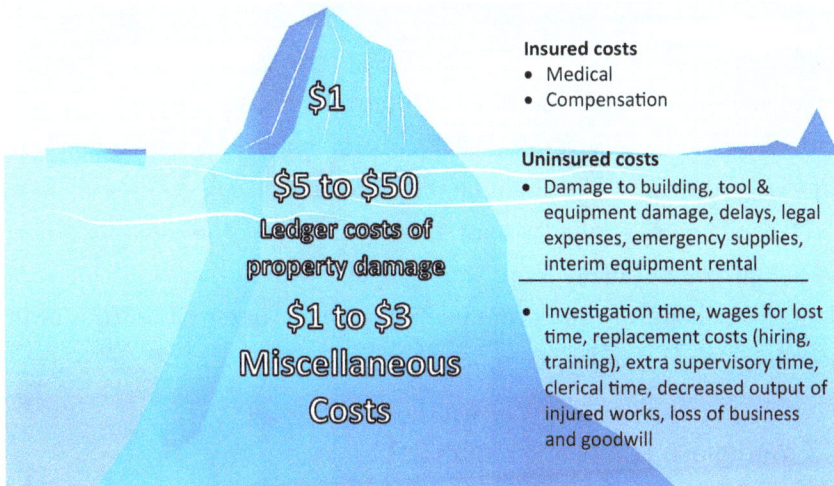

Figure 1.1   The cost of incidents

Workplace Safety and Health Institute (WSHI). Before interpreting the statistics, it is important to know that as of 2014, the WSH statistics report:

a. Included work-related traffic injuries
b. Reclassified work-related back injuries due to ergonomic risks, such as work-related musculoskeletal disorders, as occupational diseases, and
c. Expanded the number of workers to include those working in all workplaces covered under the WSH Act.

As such, the workplace injury rates for 2014 are not strictly comparable with those from previous years.

The workplace fatal injury rate for all workplaces declined from 4.9 per 100,000 employed persons in 2004 to 1.9 per 100,000 employed persons in 2015 (MOM *et al.*, 2016). However, the construction industry has remained an area of concern. In the first half of 2016, it was responsible for 43% (18) fatalities. The persistently poor safety and health performance of the construction industry made it important for all stakeholders to do their part to improve the situation.

Nevertheless, other industries such as the process and energy industries cannot be complacent because unlike the construction industry or marine industry, the process industry can cause major accidents with the extremely severe consequence of hundreds or thousands of injuries and deaths. For example, the Bhopal gas leak disaster in India (3 December 1984) killed at least 4,000. In addition, thousands of others were injured, and many more were plagued with ill health. Major accidents like Bhopal are stark reminders that major hazard industries must have high WSH standards.

1.3 Workplace Safety and Health Act

The Singapore Workplace Safety and Health Act (WSHA) is heavily influenced by the UK Health and Safety at Work, etc. Act 1974, and similar legislations in Europe and Australia. As an illustration of the performance-based approach to WSH legislation, the WSHA will be discussed herein. The WSHA was enacted in 2006 after three major accidents in 2004: the Nicoll Highway collapse on 20 April 2004, the collapse of steel latticework at the Fusionopolis Building worksite on 29 April 2004, and the fire on *Almudaina* at Keppel Shipyard on 29 May 2004. The overall intent of the WSHA was to improve the safety culture of the industry and encourage stakeholders to take reasonable practicable steps to improve WSH proactively. The WSHA is based on three basic principles:

1. Reducing risks at the source by requiring all stakeholders to eliminate or minimise the risks they create;
2. Instilling greater ownership of safety and health outcomes within the industries; and
3. Preventing accidents through higher penalties for compromises in safety management.

In contrast to the repealed Factories Act (a prescriptive legislation that details WSH requirements specifically), the WSH Act is a performance-based legislation which requires the industry to conduct risk assessment so as to proactively identify the hazards, evaluate the hazards, determine controls for the hazards, and implement and review the hazards and controls. One key feature of a performance-based approach is the concept of "as low as reasonably practicable" (ALARP). Based on the case of Public Prosecutor v Hong Jun Development Pte. Ltd [2017] SGMC 68, 'reasonably practicable' contains the following principles:

1. The risk of accident has to be weighed against the measures necessary to eliminate the risk, including the cost involved;
2. "The term 'reasonably practicable' means that stakeholders need only take preventive measures which are proportionate to the potential impact of the hazards at the workplace."
3. "...if a precaution is practicable it must be taken unless in the whole circumstances that would be unreasonable. And as men's lives may be at stake, it should not lightly be held that to take a practicable precaution is unreasonable."

In essence, "reasonably practicable" means that the benefits in reduction of risk must be weighed against the costs (e.g. time, money and resources) of implementing the controls. The weighing or evaluation of costs versus benefits should make reference to regulations, approved codes of practice (ACoP), industry standards, guidelines and norms to define what is reasonable and practicable.

Compared to the prescriptive approach in the repealed Factories Act, the performance-based regime is more sustainable because it is not possible for the government to keep creating new legislation to cover different hazards and controls. The emphasis on WSH management is an important one because penalising a company only when it has accidents or ill health is reactive and workers would already be injured or killed. Promoting effective WSH management reduces the risk of incidents and pollution. The duty holders covered in the WSH Act and their key duties are highlighted in Table 1.1.

Each duty holder has a role to play in preventing accidents and ill-health and can be taken to task for failing to perform their duties. A set

Table 1.1 Duties of duty holders under WSHA (adapted from Ministry of Manpower (2016b)

Duty Holder	Definition	Key Duties
Employer	Section 6(1) — "employer" means a person who, in the course of the person's trade, business, profession or undertaking, employs any person to do any work under a contract of service.	Section 12 — An employer must protect the safety and health of his employees or workers working under his direction, as well as persons who may be affected by their work. The employer must: • Conduct risk assessments to identify hazards and implement effective risk control measures. • Make sure the work environment is safe. • Make sure adequate safety measures are taken for any machinery, equipment, plant, article or process used at the workplace. • Develop and implement systems for dealing with emergencies. • Ensure workers are provided with sufficient instruction, training and supervision so that they can work safely.
Principal	Section 4(1) — "principal" means a person who, in connection with any trade, business, profession or undertaking carried on by him, engages any other person otherwise than under a contract of service — (a) to supply any labour for gain or reward; or (b) to do any work for gain or reward.	Section 14 — A principal must ensure that the contractor he engaged: • Is able to perform the work they are engaged for. • Has made sure that any machinery, equipment, plant, article or process that is used at work is safe. However, if the principal instructs the contractor or the workers on how the work is to be carried out,

(*Continued*)

Table 1.1   (*Continued*)

Duty Holder	Definition	Key Duties
		his duties will include the duties of an employer.
Occupier	Section 4(1) — "occupier", in relation to any premises or part of any premises, means: (a) in the case of a factory where a certificate of registration has to be obtained in relation to the premises pursuant to any regulations — the person who is, or is required to be, the holder of the certificate; (b) in the case of a factory where a notification has to be submitted in relation to the factory pursuant to any regulations — the person who is named in the notification, or is required to submit a notification; and (c) in the case of any other premises — the person who has charge, management or control of those premises either on his own account or as an agent of another person, whether or not he is also the owner of those premises;	Section 11 — An occupier must ensure that the following are safe: • The workplace. • All pathways to and from the place of work. • Machinery, equipment, plants, articles and substances. The occupier must ensure that the above does not pose a risk to anyone within his premises, even if the person is not his employee. Section 19 — The occupier may also be responsible for the common areas used by his employees and contractors. Common areas include the following: • Electric generators and motors. • Hoists and lifts, lifting gears, lifting appliances and lifting machines. • Entrances and exits. • Machinery and plants.
Manufacturer or supplier of machinery and equipment or hazardous substances	Not defined in WSHA	Section 16 — A manufacturer or supplier must ensure that any machinery and equipment or hazardous substances he provides are safe. He must:

(*Continued*)

Table 1.1   (*Continued*)

Duty Holder	Definition	Key Duties
		• Provide information on health hazards and how to safely use the machinery, equipment or hazardous substance. • Examine and test the machinery, equipment or hazardous substance to ensure that it is safe for use. • Provide results of any examinations or tests of the machinery, equipment or hazardous substances.
Installer or erector of machinery	Not defined in WSHA	Section 17 — The installer or erector of machinery must ensure that the machinery and equipment that he has erected, installed or modified is safe and without safety or health risks when properly used.
Employee	Section 6(1) — "employee" means any person employed by an employer to do any work under a contract *of* service. (Note that "contract *for* service" refers to "an independent contractor, such as a self-employed person or vendor, is engaged for a fee to carry out an assignment or project." (Ministry of Manpower 2017))	Section 15 — An employee must: • Follow the workplace safety and health system, safe work procedures or safety rules implemented at the workplace. • Not engage in any unsafe or negligent act that may endanger himself or others working around him. • Use the personal protective equipment provided to him to ensure his safety while working. He must not tamper with or misuse the equipment.
Self-employed	Not defined in WSHA	Section 13 — A self-employed person is required to take measures to ensure the safety and health of anyone in the workplace who may be affected by his work.

of subsidiary legislations are written under the WSH Act, and to support the subsidiary legislations and the WSH Act, a wide range of standards were developed. Standards approved by the Ministry of Manpower (MOM) are known as Approved Codes of Practice (ACoP) and they have higher standing in the courts as compared to non-approved codes of practice. The list of ACoPs is published in the government Gazette.

The WSH Act imposes a maximum penalty of S$500,000 for a corporate body with a first conviction. For a repeat offender, this is increased to S$1 million (MOM, 2016a). For individuals, the maximum penalty is S$200,000 for a first conviction and S$400,000 for a repeat offender. Individuals can also be imprisoned for a maximum of two years. The Ministry of Manpower can also impose composition fines instead of prosecuting. Each offence may be compounded to a sum not more than half the maximum fine prescribed for the offence or S$5,000, whichever is lower.

1.4 Overview of Workplace Safety and Health Management

Cost, quality, competition, profit, timeline, environmental pollution and Workplace Safety and Health (WSH) are different types of challenges that managers need to handle on a daily basis. On the surface, these challenges are unrelated and independent. However, the reality is that they are intertwined and inter-dependent (see Figure 1.2).

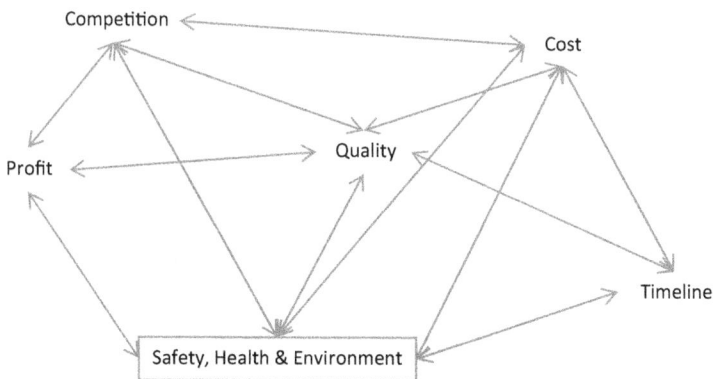

Figure 1.2   The inter-dependent nature of all management issues

For example, in order to push for project progress, a project manager may instruct workers to bypass safety and health procedures. A safety inspection may be cancelled, or the safety barricades may not be installed to save time. This leads to higher risk, and if accidents happen, the workplace will be subjected to investigations, leading to delays and loss of profit. Quality and WSH are related because when workers rush their work, not only are they more likely to make safety-related errors, they can easily make mistakes leading to quality issues. The poor workmanship can also mean rework and that means more time lost. However, WSH incidents do not occur every time WSH procedures and measures are breached. This is because there is always an element of uncertainty or luck. Thus, managers tend to focus on certain — typically, more urgent — issues that they are facing. This frequently leads to the neglect of WSH management. To counteract this tendency, it is important for managers to focus on the fundamentals of WSH management, which are aligned with the principles of good management and planning.

The three basic principles of the WSH Act show the importance of WSH management — in particular, Principle 3, which implies that companies can be penalised even if there is no accident on site. Organisations are expected to manage safety proactively and all stakeholders have to be involved. The key principles of the WSH Act are also aligned with the well-known accident pyramid or ratio (see Figure 1.3), which was developed based on a set of accident data that F. E. Bird collected (Bird, Germain

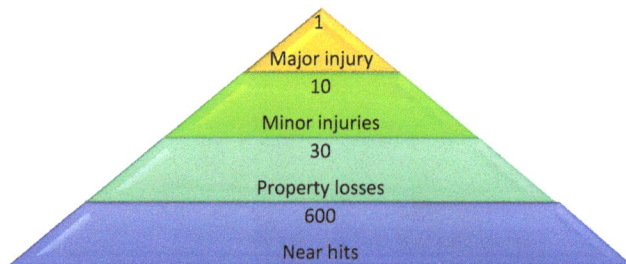

Figure 1.3   F. E. Bird's accident pyramid

and Clark, 2013). The pyramid shows that for each major injury there are disproportionately more minor injuries, property losses and near hits. This implies that if management continuously learns from near hits, property losses and minor injuries, it is possible to prevent major injuries from happening. Nevertheless, after the BP Texas oil refinery explosion in 2015, many academics warned that the concept of an accident pyramid can also be misleading. Major accidents, e.g. the major explosion of a plant and the collapse of a large structure, can have very different causes from minor injuries, property losses and near hits, especially if the nature of the incident is very individual, e.g. a worker cut his finger or fell when walking. The accident pyramid should only be taken as a guide, but its emphasis on proactively identifying incidents of low consequence to learn from is still important, as long as managers are conscious of the difference between major accidents and minor individual accidents. In addition, the ratio of 1:10:30:600 is not fixed and can change when applied to different data sets.

WSH management is based on the Plan–Do–Check–Act (PDCA) cycle (Figure 1.4), which is also adopted in quality management, environmental management and risk management. During *Plan*, organisations must proactively identify the possible hazards (WSH) or aspects (environmental) that can lead to accidents and ill health or environmental impacts. A series of controls or programmes will be identified during planning. During *Do*, the organisation will implement the controls and programmes to control the risk posed by the hazards and aspects. The effectiveness of the planning and doing processes will be *Checked*. If

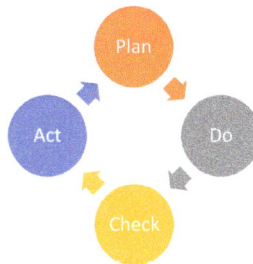

Figure 1.4   The PDCA cycle

there are gaps, the organisation will *Act* on those identified issues for continuous improvement. The cycle continues to ensure progressive improvement.

In the context of WSH management, there are several standards and some guidance available to help organisations understand the policies, structure, procedures and actions needed to maintain a management system that continually improves WSH performance. The BS OHSAS 18001 Occupational Health and Safety Management Systems Standard (British Standards Institute, 2007) is one of the most commonly adopted WSH management standards internationally. The guidance document for OHSAS 18001 is OHSAS 18002 (British Standards Institute, 2008), which contains the details in OHSAS 18001 and additional guidance and explanations. The OHSAS 18001 is similar to ISO 14001 (British Standards Institution, 2015), but the latter is designed for environmental management. Both OHSAS 18001 and ISO 14001 use an extended version of the PDCA cycle (see Figure 1.5), which includes the element of Safety, Health and Environmental (SHE) Policy. A SHE policy statement is a

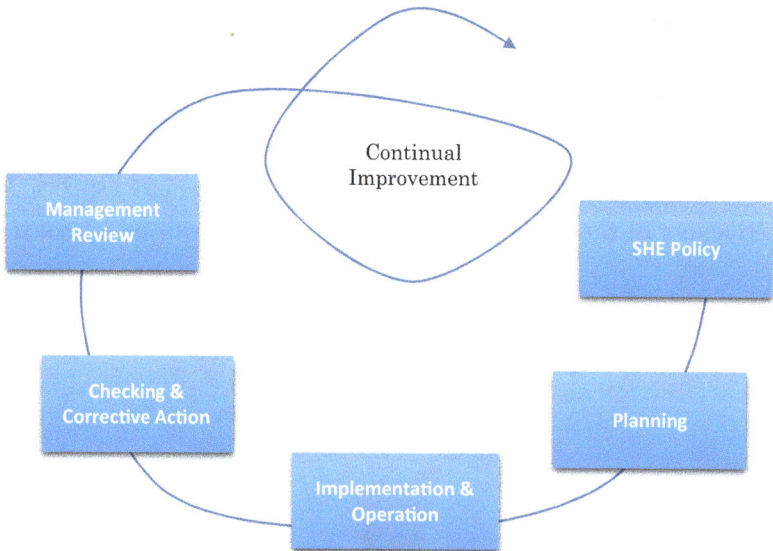

Figure 1.5   The PDCA cycle (extended)

way for the organisation and its management to express their commitment towards SHE. At the time of writing, the ISO 45001 had just replaced BS OHSAS 18001. ISO 45001 will be discussed in Chapter 6.

1.5 System Dynamics

System dynamics (also known as systems thinking) is a discipline to help managers understand inter-relationships between system components so as to better manage complex systems, which can behave unpredictably. One of the fundamental principles of system dynamics is the importance of looking beyond the direct and immediate causes of events. A systems thinker will seek to understand the patterns of behaviour of the system, system components and stakeholders so as to uncover the underlying factors and structure that caused the event to occur. This is very much aligned with the concept of root cause analysis commonly found in WSH literature. However, system dynamics takes a macro view of underlying factors and structure. It tends to go beyond the WSH management system and safety culture, to see the organisation as a whole system. This holistic view of WSH management is essential in improving WSH performance, because WSH issues may have non-WSH origins. In addition, system dynamics emphasise that complexities arise from simple dynamics such as the circular dependence between system parameters and components, and delays (Goh *et al.*, 2010). We must note, however, that system dynamics literature tend to overemphasise quantitative approaches to model these complexities through stocks and flow models. In contrast, this book adopts the key principles of system dynamics and takes a qualitative approach to their application in WSH management.

Review Questions

1. Why is Safety, Health and Environmental (SHE) management important?
2. In contrast to workplace accidents, why are occupational health and environmental pollution frequently neglected?

3. In terms of the construction industry, why is it important for all stake-holders to emphasise the importance of WSH management?
4. What are the implications of F. E. Bird's accident pyramid?
5. Describe the PDCA cycle in the context of WSH management.

References

Bird, F. E., Germain, G. L., and Clark, M. D. (2003). *Practical loss control*, Det Norske Veritas (U.S.A.), Inc., Duluth, Georgia.

British Standards Institute (2007). "BS OHSAS 18001:2007 Occupational health and safety management systems — Requirements." BSI, London.

British Standards Institute (2008). "BS OHSAS 18002:2008 Occupational health and safety management systems — Guidelines for the implementation of OHSAS 18001:2007." BSI, London.

British Standards Institution (2015). "BS EN ISO 14001:2015 Environmental management systems — Requirements with guidance for use." British Standards Institution, London.

Goh, C. L. (2008). "NZ engineer suspended from practice for two years." *Straits Times*, Singapore Press Holdings Limited, Singapore.

Goh, Y. M., Brown, H., and Spickett, J. (2010). "Applying systems thinking concepts in the analysis of major incidents and safety culture." *Safety Science*, 48, 302–309.

Goh, Y. M., and Soon, W. T. (2014). *Safety Management Lessons from Major Accident Inquiries*, Pearson, Singapore.

International Labor Organization (2009). "Facts on safety and health at work." International Labor Organization, Geneva.

Ministry of Manpower (2016a). "General penalties." <http://www.mom.gov.sg/workplace-safety-and-health/workplace-safety-and-health-act/liabilities-and-penalties>. (Nov 16, 2017).

Ministry of Manpower (2016b). "WSH Act: responsibilities of stakeholders." <http://www.mom.gov.sg/workplace-safety-and-health/workplace-safety-and-health-act/responsibilities-of-stakeholders>. (Nov 17, 2017).

Ministry of Manpower (2017). "Contract of service." <http://www.mom.gov.sg/employment-practices/contract-of-service>. (Nov 14, 2017).

Ministry of Manpower, Workplace Safety and Health Council, and Workplace Safety and Health Institute (2016). "WSH 2018 PLUS: Advancing Workplace Safety & Health In Singapore for 2018 and beyond." <http://www.mom.gov.sg/~/media/mom/documents/safety-health/publications/wsh-2018-plus.pdf?la=en> (13 June 2018).

2 Incident Causation

2.1 Definitions

Prior to the discussion on incident causation models, it is necessary to first establish the difference between accidents, incidents and events (see Figure 2.1). An event is an episode of something that happened, which may or may not have any consequences. Examples of an event include an overseas trip, a meeting and watching a movie. Accidents are unexpected and undesirable events that result in negative consequences such as injuries, fatalities, illnesses, property damage and financial loss. Accidents are a subset of incidents, which include near misses or, more accurately, near hits, which are undesirable events that could have resulted in negative consequences if the negative consequences had not been averted by chance or successful safety or emergency measures. Based on the above definitions, incidents will include workplace accidents, environmental pollution and occupational diseases.

When an event occurs, different people may view the same event differently and can derive different insights from the same event. The differences in insights can then lead to different actions with varying levels of effectiveness. The differences in perception and insights are influenced by differences in mental models, which are our internal representations of how the world works. This internal representation is made up of simplified concepts about the real world and the relationships between the concepts. Mental models are like maps, which are simplifications of the real world, but they are useful in guiding our actions. Thus, when an incident happens, it is important to have a suitable incident causation model that guides us in the way that we view

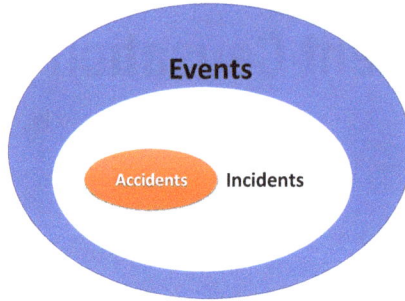

Figure 2.1   Defining accident, incident and event

and investigate accidents. A good incident causation model will help us derive suitable insights and actions to prevent future incidents.

2.2 Causation Models

There have been many incident causation models developed over the years. This section will present some of the relevant models. The Domino Theory by Heinrich (1936) (see Figure 2.2) is one of the earliest incident causation models — but because it is now considered unsuitable, it is no longer widely used. It focuses on how the unsafe acts, carelessness and anti-social behaviours (e.g., alcoholism) of individuals contribute to accidents. According to the Domino Theory, the fundamental cause of an accident, as represented by the last block of the domino, is the "social environment and inherited behaviour" of the victim and/ or workers responsible for the accident. The Domino Theory reflects the prevalent mental models of the past, which were usually focused on the equipment, environmental factors and human error (including violations). Such a mental model is still commonly found. For example, in July 2017, after the collapse of a viaduct that was under construction in Singapore, the media (Christopher Tan, 2017) highlighted human error as the cause of the accident. Blaming fault of person, social environment and inherited behaviour (non-management related factors) as the fundamental cause of incidents can lead to management

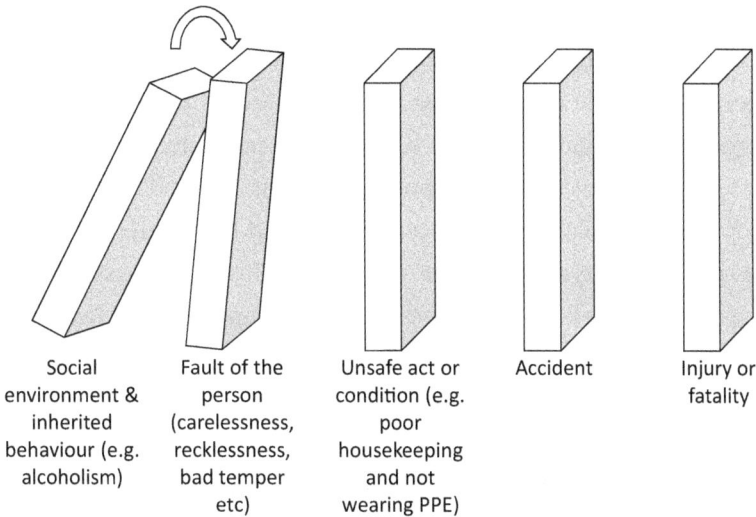

| Social environment & inherited behaviour (e.g. alcoholism) | Fault of the person (carelessness, recklessness, bad temper etc) | Unsafe act or condition (e.g. poor housekeeping and not wearing PPE) | Accident | Injury or fatality |

Figure 2.2   An adapted version of Heinrich's Domino Theory; "PPE" in the figure refers to Personal Protective Equipment

blaming the workers for incidents and ending investigations before more fundamental or underlying factors are uncovered. The failure to remove or mitigate the underlying factors can then lead to the recurrence of incidents. However, it must be noted that the individuals are also accountable for their own actions and that it is important for workers to cooperate with their employers and comply with workplace safety and health (WSH) rules and regulations.

The Domino Theory is now considered unsuitable because organisations have come to realise that there are other more fundamental underlying factors contributing to human error, unsafe acts and unsafe conditions. Some of these underlying factors include poorly designed procedures, production pressure, error-provoking equipment design and lack of management commitment to safety. The general agreement among modern incident causation models is that, even though incidents are caused by unsafe equipment, environment and/or behaviours, organisations are expected to manage these direct causes in a proactive and *systemic* manner.

Incident causation models that highlight the role of the organisation, its policies, procedures, structure and culture in the causation of an incident are classified as *systemic models*. In fact, there are also models that go beyond organisations to consider the influence of the industry, government and even beyond, but this book is focused on the prevention of incidents at the organisational level and the influence of factors beyond the organisation are not discussed. Numerous systemic models have been developed over the years, for example, the Management Oversight Risk Tree (MORT) (Johnson, 1980), the contributing factors in accident causation (CFAC) model (Sanders, 1988), the Swiss Cheese Model (SCM) (Reason, 1997), the Loss Causation Model (LCM) (Bird *et al.*, 2003), and the Modified Loss Causation Model (MLCM) (Chua and Goh, 2004). The Event Causation Technique (ECT), which is an extension of the MLCM, is the main systemic incident causation model that will be discussed, and it will be presented in detail in the following chapter. These models implicitly or explicitly reinforce the concept of multiple-causation, where the cause of an incident does not lie in a single line of causation, but often branches into various chains of factors.

Figure 2.3 captures the gist of systemic models. Events, as explained earlier, can be any episode of "things happening", including incidents. "Patterns" refer to patterns of events over time or patterns across entities like people, machines or organisations. Finally, "underlying factors" refer to management system (e.g. ISO45001 and ISO 14001) weaknesses, leadership inadequacies or organisational culture factors. The pyramid in Figure 2.3 indicates that events and incidents do not "just happen"; they arise due to certain systematic patterns of

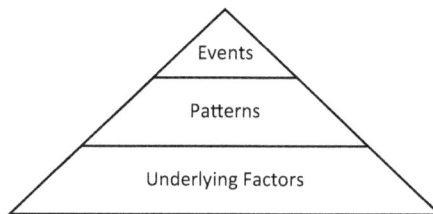

Figure 2.3   Systemic view of events; adapted from Senge (2006)

the way things are done in the organisation or certain characteristics of the organisation. Some examples of patterns can be the declining number of and competency of engineers across time, and the pattern of unsafe behaviour (e.g. not putting on their harness and lanyards) across workers. These observable patterns are usually related to poor management and organisational culture. Poor management refers to a weak management system, which does not have an effective plan–do–check–act (PDCA) cycle. For example, an organisation that does not have a systematic procedure for conducting risk assessment and this was not identified during periodic management system audits. The lack of risk assessment then led to hazards not being assessed and no control being implemented to reduce the risk of the hazards. This results in high risk hazards like open edges being frequently observed on site. Finally, a worker falls from height (an event) when working near one of the open edges.

2.3 Haddon's Energy Transfer Model

Even though the Energy Transfer Model (ETM) (Haddon Jr, 1973) is not a systemic model, it provides a practical understanding of incidents and control measures. The ETM (see Figure 2.4) traces the interactions between an energy source or hazardous substance and a person or property and the resulting consequences. Ten counter-measures or controls are identified based on the model.

The ten incident controls identified by Haddon Jr (1973) are:

1. To prevent the initial marshalling of the form of energy
2. To reduce the amount of energy marshalled
3. To prevent the release of the energy
4. To modify the rate of spatial distribution of release of energy from its source
5. To separate in space or time the energy being released from the susceptible structure

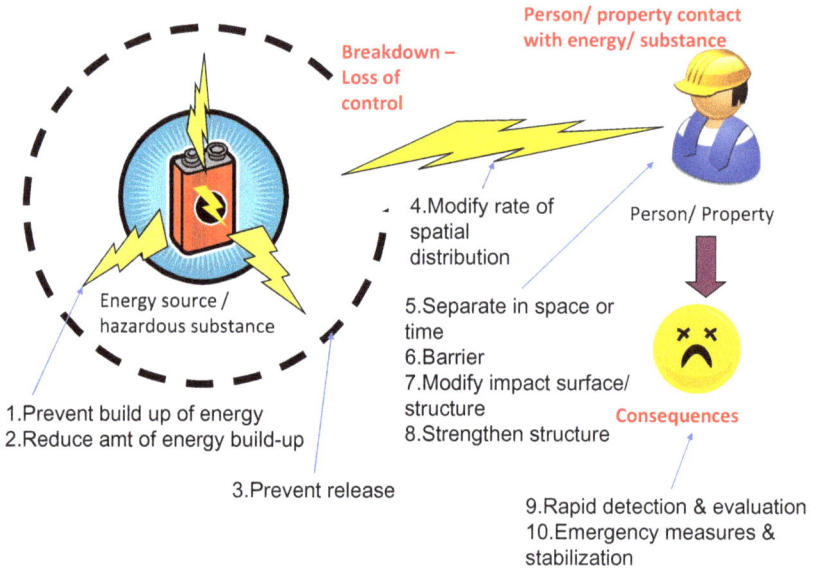

Figure 2.4   Adapted from Haddon's Energy Transfer Model (1973)

6. To separate the energy being released from the susceptible structure by interposition of a material barrier

7. To modify the contact surface, subsurface, or basic structure which can be impacted

8. To strengthen the structure which might be damaged by the energy transfer

9. To move rapidly in detection and evaluation of damage and to counter its continuation and extension

10. All those measures which fall between the emergency period following the damaging energy exchange and the final stabilisation of the process

The abbreviated forms of the ten controls are inserted into Figure 2.4. The ETM is very useful in risk assessment and incident investigation. In fact, many of the incident causation models incorporate some of the concepts highlighted in the ETM.

2.4 Loss Causation Model

The Loss Causation Model (LCM) (Bird *et al.*, 2003) has a similar structure to the Domino Theory, but it represents an important shift in mental models. With reference to Figure 2.5, the key difference between the Domino Theory and the LCM is that "lack of management control" is deemed to be the most fundamental cause of incidents and losses in the LCM. Coming from the left of the model, lack of management control (inadequate management system, inadequate standard or inadequate compliance to the system and standard) can lead to basic causes of incidents. As in the case of the Domino Theory, basic causes or root causes refer to personal factors and job or system factors. The former is defined as physiological and psychological factors, while job or system factors refer to factors such as leadership, work standards, risk assessment and maintenance.

Incidents are directly caused by immediate causes, which include substandard acts and substandard conditions. "Acts" refer to observable human behaviours, and "conditions" refer to physical conditions such as condition of equipment, material, structure and environment. An act or condition is considered substandard when it fails to meet a stipulated standard set by the organisation. This implies that organisations

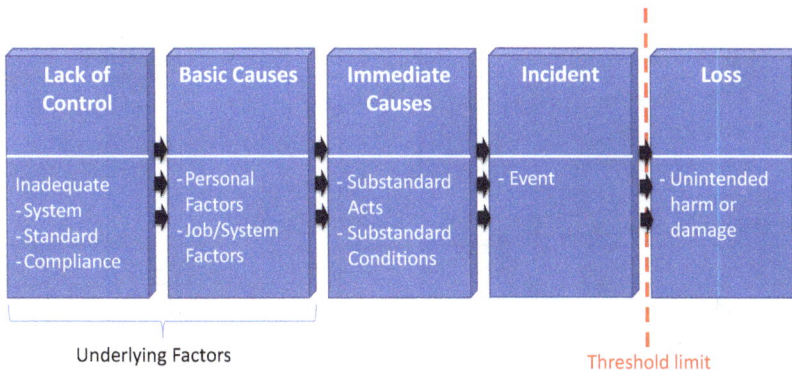

Lack of Control	Basic Causes	Immediate Causes	Incident	Loss
Inadequate - System - Standard - Compliance	- Personal Factors - Job/System Factors	- Substandard Acts - Substandard Conditions	- Event	- Unintended harm or damage

Underlying Factors

Threshold limit

Figure 2.5 Loss Causation Model; adapted from Bird *et al.* (2003)

are expected to set comprehensive standards to cover behaviours and workplace conditions. Immediate causes are generally observable and they are more easily determined as compared to basic causes.

Loss has a broader definition than human injuries and fatalities. Losses include property loss, reputation loss and environmental impact. Following Haddon's ETM, losses arise from incidents when a certain protective threshold has been exceeded. If a person is struck by a moving object, for example a forklift travelling at extremely slow speed, the collision may not exceed the person's capacity to absorb the kinetic energy exerted by the forklift. In such a situation, the incident would remain as a near-hit because the injury threshold of the person has not been exceeded and there would not be any loss.

To better illustrate the LCM, Table 2.1 captures the possible classifications under each of the main blocks in the LCM. It must be noted that these classifications are not comprehensive, and are written in a generic fashion. They are useful prompters to help people consider possible causes and factors during incident investigation and classification, but incident specific information must accompany the classification to provide the background information needed for someone to understand the reason for the classification.

Table 2.1   Loss causation model; adapted from Bird *et al.* (2003)

Losses from Incidents		
• Injured worker time	• Co-worker time	• Leader time
• General losses (lost production time, decreased effectiveness of employees and loss of business goodwill, etc.)	• Property losses	• Other losses
Event/ Incident Types		
• Struck against (running or bumping into)	• Struck by (hit by moving object)	• Fall to lower level (either the body falls or the object falls and hits the body)

(Continued)

Table 2.1 *(Continued)*

• Fall on same level (slip and fall, tip over)	• Caught in (pinch and nip points)	• Caught on (snagged, hung)
• Overstress or overexertion or overload	• Caught between (crushed or amputated)	• Contact with (any harmful energy or substance)
• Release of (any harmful energy or substance)		

Immediate Causes		
Substandard Acts or Practices		
• Operating equipment without authority	• Failure to warn	• Failure to secure
• Operating at improper speed	• Making safety devices inoperable	• Using defective equipment
• Using equipment improperly	• Failing to use personal protective equipment properly	• Improper loading
• Improper placement	• Improper lifting	• Improper position for task
• Servicing equipment in operation	• Horseplay	• Under influence of alcohol and/or other drugs
• Failure to follow procedure or policy or practice	• Failure to identify hazard or risk	• Failure to check or monitor
• Failure to react/ correct	• Failure to communicate or coordinate	•
Substandard Conditions		
• Inadequate guards or barriers	• Inadequate or improper protective equipment	• Defective tools, equipment or materials
• Congestion or restricted action	• Inadequate warning systems	• Fire or explosion hazards
• Poor housekeeping; disorder	• Hazardous environmental conditions: gases, dusts, smokes, fumes, vapours	• Noise exposure
• Radiation exposure	• Temperature extremes	• Inadequate or excess illumination

(Continued)

Table 2.1   (*Continued*)

• Inadequate ventilation	• Inadequate instructions or procedures	
Basic Causes		
Personal Factors		
• Inadequate physical or physiological capability	• Inadequate mental/ psychological capability	• Physical or physiological stress
• Mental or psychological stress	• Lack of knowledge	• Lack of skill
• Improper motivation		
Job/System Factors		
• Inadequate leadership and/or supervision	• Inadequate engineering	• Inadequate purchasing
• Inadequate maintenance	• Inadequate tools and equipment	• Inadequate work standards
• Wear and tear	• Abuse or misuse	
Lack of Control (System Elements)		
• Leadership and Administration	• Leadership training	• Planned inspections and maintenance
• Critical task analysis	• Incident investigation	• Performance observation
• Emergency preparedness	• Rules and work permits	• Incident analysis
• Knowledge and skill training	• Personal protective equipment	• Health and hygiene control
• System evaluation	• Engineering and change management	• Personal communications
• Team communications	• General promotion	• Hiring and placement
• Materials and services management	• Off-the-job safety	

2.5 Swiss Cheese Model

The Swiss Cheese Model (SCM) (Reason, 1997) shown in Figure 2.6 is used in many industries, such as the oil and gas, aviation, mining and nuclear industries. Based on the model, consequences are the result of several conditions: (1) the presence of a threat or danger that emits an

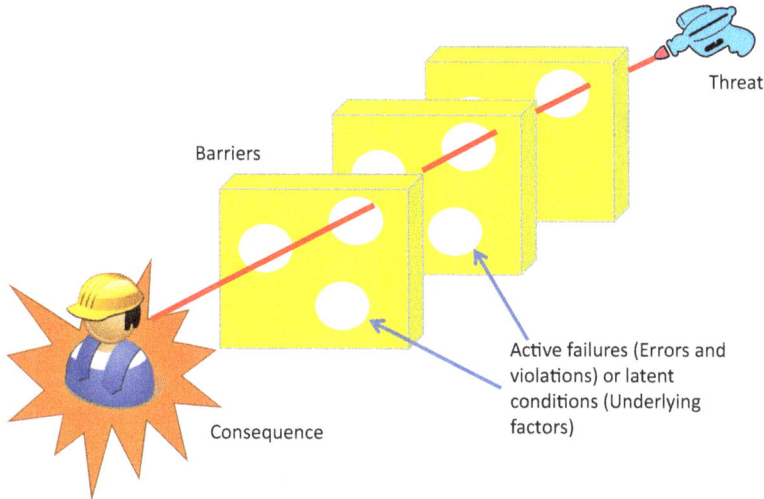

Figure 2.6   Swiss Cheese Model; adapted from Reason (1997)

accident trajectory, represented by the laser beam; (2) the presence of holes in the barriers created by active failures and/or latent conditions; and (3) the alignment of the holes to allow the accident trajectory to hit the 'target', such as a property, equipment or person. The barriers are essentially synonymous with the controls in the LCM and active failures are essentially the immediate causes. Latent conditions are like basic causes in the LCM. According to Reason (1997), active failures are the errors and violations committed at the 'sharp end' of the system; that means by frontline workers, pilots, maintenance personnel and operators. Active failures have relatively immediate or short-lived effects. Latent conditions are like pathogens to the human body, e.g. poor design, gaps in supervision, undetected manufacturing defects or maintenance failures, unworkable procedures, clumsy automation, shortfalls in training, inadequate tools and equipment. These latent conditions can be present in the organisations many years before they interact with active failures and local situations to defeat the different layers of barriers in the organisation. These latent conditions originate from high level decisions of managers, designers, manufacturers, regulators and governments.

Even though there are differences between the LCM and the SCM, both models place emphasis on management control. The key message is that incidents and losses will occur if there is a lack of management control. However, Reason (1997) highlighted safety culture as an important aspect of latent conditions, which the LCM did not cover explicitly. The concept of safety culture will be covered in Chapter 7. The SCM also emphasised that the holes in the barriers are inevitable, so the key is to reduce these holes as much as possible and to reduce the likelihood of the holes being aligned.

2.6 Causal Loop Diagrams

Causal loop diagrams (CLDs) are one of the tools used in system dynamics (or systems thinking) to describe the complex underlying factors influencing system behaviours (Senge, 2006). Goh *et al.* (2010) introduced the use of CLDs in analysing incidents. In contrast to the linear nature of most incident causation models, a CLD can be used to describe the circular nature of cause and effect, and explain the system behaviour over time. A CLD is made up of three basic processes: a reinforcing feedback loop, balancing feedback and delays. A reinforcing loop exists where a behaviour encourages similar behaviour in the future. It amplifies the behaviour across time and as the reinforcing loop continues, an accelerating growth or decline will occur. Figure 2.7 is an example of a reinforcing loop where the safe behaviour and its positive consequences, such as recognition by a supervisor and incentives from the company, creates a virtuous cycle that encourages growth in the safe behaviour in the future, which then leads to further safe behaviour and positive consequences. A reinforcing loop can also be a vicious cycle where the safe behaviour results in negative consequences like disapproval from peers and complaints from "unenlightened" clients who do not tolerate slight delays due to additional safety checks. These negative consequences will lead to less safe behaviour in the future.

Balancing feedback loops describe processes that aim to balance a behaviour or indicator at a target level. For example, as depicted

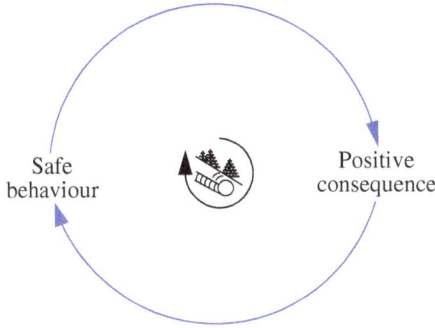

Figure 2.7   Example of a reinforcing loop

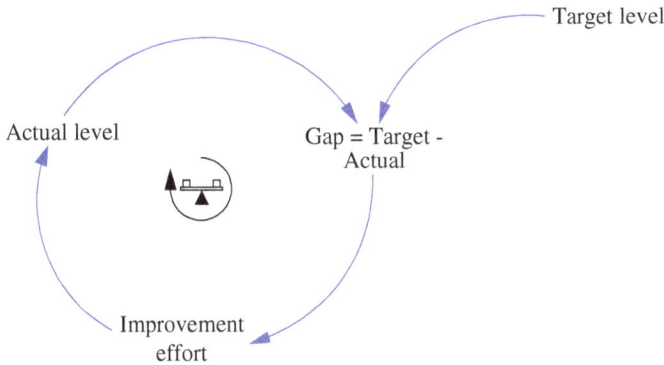

Figure 2.8   Example of a balancing loop

in Figure 2.8, a company may be monitoring a safety performance indicator ("Actual level") and the company is motivated to increase its improvement effort whenever there is a gap between its target level and actual level. The larger the size of the gap, the greater the effort to improve the situation. When the size of the gap reduces, the pressure to improve the safety performance reduces. The way to sustain the improvement effort is to adjust the target level higher to maintain a healthy gap between target and actual levels. The target level need not be explicit and can be implicit and hidden in people's minds. It is also possible for both explicit and implicit target levels to exist, and people may unknowingly be focused on the implicit target despite the explicit target level being higher.

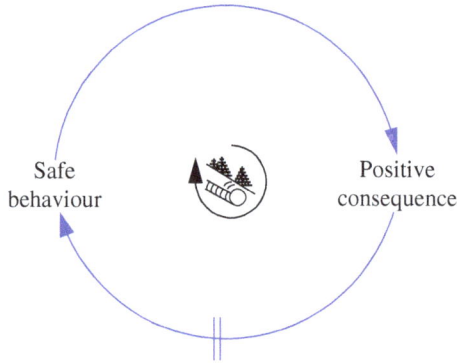

Figure 2.9   Example of a delay in a reinforcing loop

The subsequent effect of a cause may be delayed. With reference to Figure 2.9, when a safety behaviour leads to a positive consequence, it might take some time for workers to be convinced that the safe behaviour can indeed lead to positive consequences. The delay is represented by two parallel lines on the arrow linking two variables. A lack of awareness of these delays can cause improvement efforts and safety programmes to be aborted prematurely.

Using the basic building blocks of reinforcing loops, balancing loops and delays, a series of systems archetypes can be captured to highlight common organisational problems and identify the possible leverage points that organisations can focus on to improve organisational performance. The systems archetypes are useful for understanding the problems inhibiting WSH management and provide suggestions on how to improve WSH performance. These archetypes are described in Senge (2006) and can be easily found on the internet. The following will describe the "shift the burden" archetype (see Figure 2.10) as an illustration of system archetypes.

The archetype consists of two balancing loops, one addressing the symptoms, the other the underlying or fundamental causes of the problem. Symptomatic "solutions" can be very attractive to management, producing relatively quick positive results that focus on the symptoms and relieve the immediate pressure of the problem. The second balancing process focuses on fundamental solutions to the problem that are more sustain-

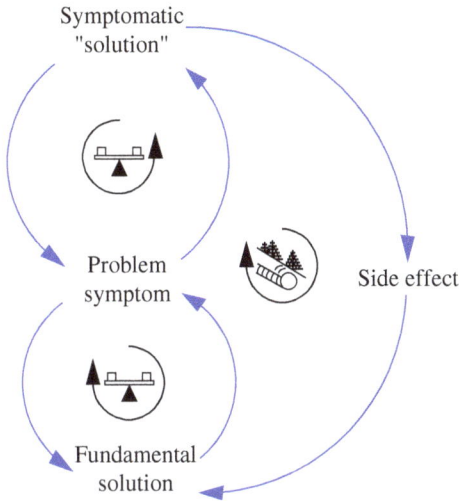

Figure 2.10   Shifting the burden archetype

able but also have a delay between implementation and results. Failure to recognise this delay contributes to the tendency to focus on the symptomatic solution. As resources are focused on the symptomatic solution, the fundamental problem and the capacity of the organisation to resolve the problem worsens (side effect). Systems thinking concludes that short-term efforts aimed at the symptoms can be useful, particularly to relieve the pressure as the delay from the fundamental solution is experienced, but long-term solutions to a problem must focus on the fundamental loop.

Archetypes help managers understand the possible underlying factors sustaining the pattern of system behaviour, which leads to the occurrence of the incident. The archetypes also present leverage points and possible strategies that managers can use to improve their current situation. The archetypes and CLDs also allow managers to present their mental models to allow others to examine them and discuss solutions.

2.7 Conclusions

This chapter presented several incident causation models, and it was established that systemic models are now widely adopted by organisations and the workplace safety and health (WSH) industry. Systemic models emphasise that organisational factors like the management system, leadership and safety culture are the fundamental causes of workplace incidents. These organisational factors can be described and evaluated using causal loop diagrams (CLDs). CLDs can be used to describe system archetypes, which are generalised descriptions of underlying factors influencing system behaviour.

Review Questions

1. Describe the difference between an "event", an "incident" and an "accident".
2. Why is the Domino Theory no longer accepted in modern WSH management?
3. Give examples of events, patterns and underlying factors. Demonstrate how the underlying factors cause the patterns and then the events.
4. What is the key difference between a systemic model (e.g. Swiss Cheese Model) and non-systemic model (e.g. Energy Transfer Model)?
5. Explain the difference between immediate cause, basic cause and lack of control.
6. Explain the similarities and differences between the Swiss Cheese Model and the Loss Causation Model.
7. Do a search on the internet for the different system archetypes. Explain how to select suitable archetypes for different system problems.

References

Bird, F. E., Germain, G. L., and Clark, M. D. (2003). *Practical loss control leadership*, Det Norske Veritas (U.S.A.), Inc., Duluth, Georgia.

Christopher Tan (2017). "PIE uncompleted highway structure collapse: Cause likely to be human error, says veteran engineer." *Straits Times*, SPH, Singapore.

Chua, D. K. H., and Goh, Y. M. (2004). "Incident causation model for improving feedback of safety knowledge." *J. Constr. Eng. and Manage. — Am. Soc. of Civ. Eng.*, 130(4), 542–551.

Goh, Y. M., Brown, H., and Spickett, J. (2010). "Applying systems thinking concepts in the analysis of major incidents and safety culture." *Safety Science*, 48, 302–309.

Haddon Jr, W. (1973). "Energy damage and the ten countermeasure strategies." *Human Factors*, 15(4), 355–366.

Heinrich, H. W. (1936). *Industrial Accident Prevention*, McGraw Hill, New York.

Johnson, W. G. (1980). *MORT Safety Assurance System*, Marcel Dekker, New York.

Reason, J. (1997). *Managing the risks of organizational accidents*, Ashgate, Aldershot.

Sanders, M. S., and Shaw, B. (1988). *Research to determine the contribution of system factors in the occurrence of underground injury accidents*, Bureau of Mines, Pittsburgh.

Senge, P. (2006). *The fifth discipline — The art & practice of the learning organisation*, Random House Australia, New South Wales.

3 Incident Investigation

3.1 Introduction

This chapter will provide an overview of incident investigation. The intent is not to provide in-depth information, which is not necessary for most readers. The chapter will cover the purpose of incident investigation, an overview of the investigation process, types of evidence, incident analysis and the event causation technique (ECT).

3.2 Purpose of Investigation

According to ISO 45001:2018 clause 10.2 Incident, nonconformity and corrective action, organisations must "establish, implement and maintain a process(es), including reporting, investigating and taking action, to determine and manage incidents and nonconformities". In general, the aims of an investigation are to:

1. determine root causes (e.g. latent conditions and basic causes) and other factors that might be causing or contributing to the occurrence of incidents;
2. determine if similar incidents have occurred or could potentially occur;
3. review existing risk assessment and determine and implement any action needed, including corrective actions, in accordance to the hierarchy of controls and the management of change;
4. recommend changes to the WSH management changes, if necessary; and
5. communicate the investigation results.

Note that corrective actions are actions to eliminate the cause of a detected nonconformity (i.e. non-fulfilment of a requirement) or other undesirable situation, and to prevent recurrence. Previously, OHSAS 18001:2007 also indicated the need for preventive actions and continual improvements to be identified during investigation. Preventive actions are actions to eliminate the cause of a *potential* nonconformity or other undesirable *potential* situation. Continual improvement refers to implementation of the PDCA cycle, in particular the "check" and "act" portions. See Case Study 1 for examples of WSH deficiencies, corrective action, preventive action, and continual improvement.

Case Study 1 — Forklift Accident
A worker was struck by a forklift at a worksite leading to permanent injuries to the legs of the injured worker. The investigation found that the forklift driver was driving the forklift forward while carrying a load, which blocked the driver's view. The safe method was to reverse the forklift to ensure that there was a clear view of the path and any pedestrians in the vicinity. During the investigation it was discovered that the worker who drove the forklift did not attend suitable forklift training and had driven the forklift without permission. However, the forklift key was on the forklift and was not placed in the key press as per procedure. A key press is a small cupboard meant to control access to keys through a log book system. There was no clear requirement on how frequently the supervisor, manager and workplace safety and health officer (WSHO) overseeing the key press should check on the keys in the key press.
Causes: Blocked vision when driving forklift ← unsafe driving ← untrained forklift driver ← uncontrolled access to forklift keys ← key not placed in key press ← inadequate monitoring and control of key press
Corrective actions for "inadequate monitoring and control of key press"   1. Develop a procedure for control of the key press, where all drivers must return the keys to the key press whenever the forklifts are not in use. Supervisors, workplace safety and health officers (WSHO) and managers have to conduct different levels of checks as stipulated in the new procedure.   Corrective actions for "key not placed in key press"   2. Make missing keys obvious with good visual markers in the key press.   3. Drivers will be required to place an ID at the key press in exchange for the key.   4. All forklift drivers are briefed on the importance of removing the keys from the forklift and returning the keys to the key press. Disciplinary action will be taken

(Continued)

Case Study 1 *(Continued)*

---

if drivers were found to flout the safety rule. Incentives will be provided for drivers that comply with the procedure consistently.

Corrective actions for "untrained forklift driver"

5. All workers are reminded that unauthorised use of the forklift is dangerous and will be punished.
6. Authorised forklift drivers will be identified with a brown helmet with a name on it. Pictures of authorised forklift drivers are displayed on the notice board.

Preventive actions arising from this accident

1. The preventive maintenance of the forklift will be implemented.
2. Additional pedestrian pathways will be created to reduce the likelihood of collision.

Actions for continual improvement arising from this accident

1. The effectiveness of the measures will be reviewed by the safety committee 1, 3 and 6 months after implementation. Subsequently, a 6-monthly review will be conducted as part of the 6-monthly audit.

The key purpose of an investigation is to prevent a recurrence and improve the management system and culture by eliminating or mitigating the systematic patterns of behaviour and management that produced the incident. As discussed in Chapter 2, incidents can be very costly. Thus, it is important that investigations go deep enough to identify patterns and underlying factors and not stay on the event level.

Figure 3.1 demonstrates how incident investigation fits into the overall picture of WSH management. Incident investigation is a reactive component of a management system. It is reactive in the sense that investigation is triggered only when there is an incident, which is undesirable due to its consequences or potential consequences. Nevertheless, incident investigation serves an important role in improving the way operations are being managed and controlled. On the other hand, it will be ideal if hazards and risk controls are identified proactively so that operations are safe and incident-free. The key underlying process in WSH management that promotes proactive management is risk assessment. Risk assessments are conducted prior to operations, at scheduled

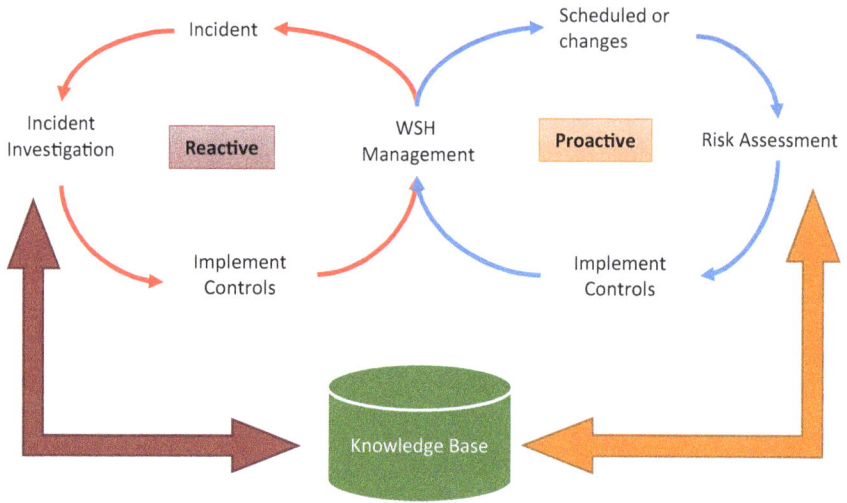

Figure 3.1 Dependence between reactive and proactive control loops; adapted from Chua and Goh (2004)

periods, after an incident and when there are changes in the operations (e.g. method, equipment and material). As depicted in Figure 3.1, both proactive and reactive loops aim to improve the WSH performance of the operation.

To ensure that the organisation learns from its past experiences, it is important to capture the information generated during incident investigation and risk assessment in a knowledge base, so that investigators and risk assessment (or risk management) teams can access the knowledge base when they are conducting investigations and risk assessments respectively. For example, during an investigation, it is important to consider the risk controls that were identified in the risk assessment. The investigators should assess if the risk controls were comprehensive, suitable, and implemented. In this way, the risk assessment process can also be reviewed as part of the incident investigation process. An incident investigation should make full use of the undesired incident to improve the underlying structures and factors influencing the patterns of how operations are being conducted in the organisation. If not, the organisation would have missed an opportunity to learn.

3.3 Overview of Investigation Process

An investigation can be split into the four inter-related phases: notify, investigate, analyse and report. Organisations need to be notified of incidents before an investigation can be conducted. During notification, the organisation needs to prioritise and allocate resources to the investigation. Some considerations include the severity of the incident, the likelihood of a recurrence, the worst credible consequence of the incident, and the resources available. Many organisations launch investigations only when the incident severity is high, e.g. when a worker is hospitalised or severely injured. However, the actual severity of an incident is frequently a matter of luck. A worker struck by a falling wrench could have easily survived if the wrench fell one second later, and that will reduce the incident consequence to zero. Thus, organisations should assess the worst credible consequence of an incident, and conduct an investigation if the severity of the worst credible consequence is high. It is also important for the organisation to communicate with key stakeholders, e.g. government agencies, clients, workers and unions, on the investigation's purpose and process so as to allay concerns and garner support for the investigation.

The investigation phase is an evidence-gathering process which focuses on obtaining the facts and information related to the incident. The fact-gathering process is intertwined with the "analyse" phase, which includes hypothesising about what happened, causes, and underlying factors. Analysis aims to identify the deep-rooted causes of the incident so as to prevent the recurrence of incidents. The analysis process can involve the use of different methods to guide investigators to probe into possible causes. Missing facts and evidence will be identified during the analysis and hence guide the investigation. Finally, reporting is about presenting the findings and recommendations to management and other stakeholders. A report is usually written, and subject to management's comments and amendments.

It is important to set out the investigation policies and procedures before incidents happen. This includes developing the notification process and drawing up forms, identifying investigators for different types of accidents, deciding on the competencies required for different investigations, and allocating resources for investigators (e.g. investigation kit, financial resources, and time). It is useful to involve operations staff in investigations as they are the most familiar with the operations. Workplace safety and health professionals who are more familiar with the WSH legislations and investigation processes can assist the operational staff with and advise them on the investigation's purposes and processes. These roles and responsibilities must be determined as part of the investigation procedure.

It should be noted that the investigation process is not an exact science. Many-a-times, the investigation team will need to use their judgement and opinions to assess the facts and evidence gathered. However, it must be clearly stated when something is an opinion or a judgement, and the team should not mislead the readers into thinking that these opinions or judgements are factual. In addition, it is important for investigators to realise that they are an observer of an event that happened in the past. This retrospective view allows the investigator to see the hazards and the consequences of decisions more easily. However, the "actors" or people involved in the incident were looking forward as the events unfolded. Furthermore, the actors are busy people who are always juggling safety, productivity, quality and other issues including personal problems in life. The actors do not have full information about their work environment, and are also uncertain about the consequences of their actions and other people's actions. According to Dekker (2017), this is similar to walking in a tunnel, which limits the view of each actor. The individual characteristics of these actors, like mindfulness, diligence and capability, are important factors, but investigators need to look beyond the individuals to consider the reasons for the mistakes and errors that individuals made. Knowing the reasons for human errors is critical in preventing a recurrence because a different person placed in the same situation can easily make the same error.

For example, in Case Study 1, the unauthorised forklift operator who took the key without authorisation was probably balancing different work pressures while trying to make the best use of the time and resources that he was given. Even though it is still necessary to provide deterrence (i.e. punishment) for workers who commit such violations, it is more effective to prevent unauthorised access to the keys and reduce the likelihood for such violations. In contrast, managers can blame the operator for violating the safety rules, punish him and assume that other workers will not violate said safety rules again after witnessing a worker being punished. Most of the time, such punishment will result in fear in the organisation and can cause workers to stop providing WSH-related information to management. However, this does not mean that a "no-blame" policy will work. There is a need for fair punishment, which takes into account both the actual challenges that individuals face, and focuses on developing systems that prevent mistakes and violations.

3.4 Types of Evidence

There are four main types of evidence: part, position, people and paper. "Part" refers to evidence in the form of material, equipment and parts of the environment such as the worn-out tyres of a forklift, the ignition key of a motor vehicle, the dust level or heaps of dust, the noise level, the illumination level, the braces of scaffolds, and lanyards. Part evidence is usually collected from the incident scene. One of the first actions of an investigation is to cordon off the incident scene to prevent part evidence from being removed or altered. Position refers to the physical relationship between the part(s) and people involved, and the placement of the evidence at different points of time. Some examples include the position of the injured before, during and after the incident, the location of the safety data sheet (for chemicals or hazardous substances) before and during the accident, the position of the wheel of the truck after the explosion, and the position of the crane after the collapse.

People evidence is self-explanatory. Some examples include interviews with the injured, witnesses (people that saw the incident), people associated with the activity and people with information relevant to the investigation. Besides direct eye-witnesses and people directly involved in the activity, investigators frequently need to interview suppliers of machinery, experts or experienced workers who have a good understanding of how work is supposed to be conducted, and managers who are accountable for the work area in general. Interviews with witnesses and people with direct information about the incident should be conducted as early as possible because people's memories tend to distort with time. In addition, witnesses might become affected by other people's intentional or unintentional comments. Investigators must also obtain "papers" and documents (including electronic and video footage) related to the incident. These documents include, for example, standard operating procedures or method statements, job orders, emails, close circuit TV footage, safety data sheets (SDS), safety committee meeting minutes, shift change log books, permits-to-work, and risk assessments.

The investigation team will have to decide what evidence needs to be collected. This is usually based on investigation procedure, experience, the hypotheses of the incident, and the analyses that the investigators conduct iteratively. Each cause should ideally be based on two or more pieces of evidence.

3.5 Event Causation Technique

There are many incident analysis methods available in the literature. The Energy Institute (2008) captured 28 incident analysis methods that are suitable for analysing underlying (human and organisational) factors, but many methods were still excluded from the document. Most established incident analysis methods are useful, but their degree of usefulness really depends on how the investigator makes use of the method. Even though there are differences between the incident analysis methods, they also have significant similarities. This book introduces the Event Causation Technique (ECT).

The ECT is depicted in Figure 3.2 and Table 3.1 shows the corresponding taxonomy, which is not exhaustive. The ECT is based on the PhD dissertation of the author, but the technique was refined across the years after applications in actual accident investigations and

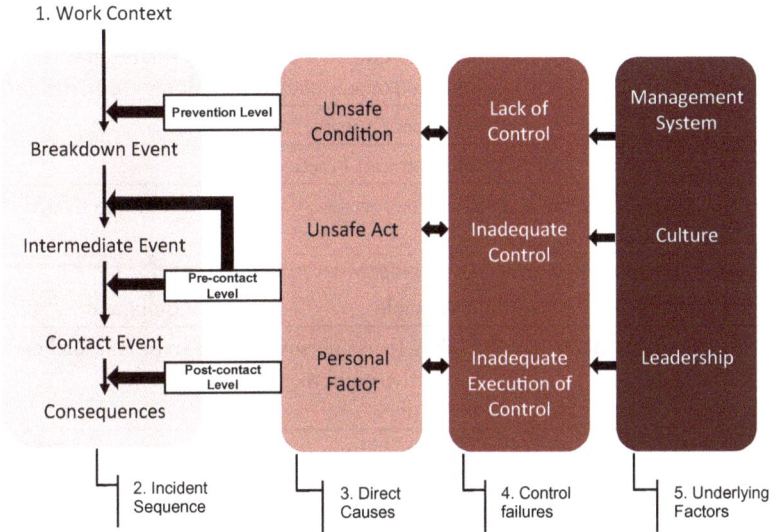

Figure 3.2   Event causation technique version 2

Table 3.1   Event causation technique taxonomy

Consequences		
1. Injured worker time	2. Co-worker time	3. Leader time
4. General losses (lost production time, decreased effectiveness of employees and loss of business goodwill, etc.)	5. Property losses	6. Other consequences
Incident Events — Breakdown, intermediate and contact events		
1. Collapse of structure	2. Loss of balance	3. Release of (any harmful energy or substance)
4. Loss control of plant/ equipment	5. Overstress/ overexertion/ overload	6. Fall to lower level (either the body falls, or the object falls and hits the body)

(*Continued*)

Table 3.1   (*Continued*)

7. Fall on same level (slip and fall, tip over)	8. Fall on same level (slip and fall, tip over)	9. Struck by (hit by moving object)
10. Struck against (running or bumping into)	11. Contact with (any harmful energy or substance)	12. Caught in (pinch and nip points)
13. Caught on (snagged, hung)	14. Caught between (crushed or amputated)	15. Other events
Direct Causes		
Unsafe Acts and Practices		
1. Operating equipment without authority	2. Failure to warn	3. Failure to secure
4. Operating at improper speed	5. Making safety devices inoperable	6. Using defective equipment
7. Using equipment improperly	8. Failing to use personal protective equipment properly	9. Improper loading
10. Improper placement	11. Improper lifting	12. Improper position for task
13. Servicing equipment in operation	14. Horseplay	15. Under influence of alcohol and/or other drugs
16. Failure to follow procedure/policy/ practice	17. Failure to identify hazard/risk	18. Failure to check/ monitor
19. Failure to react/ correct	20. Failure to communicate/ coordinate	
Unsafe Conditions		
1. Inadequate guards or barriers	2. Inadequate or improper protective equipment	3. Defective tools, equipment or materials
4. Congestion or restricted action	5. Inadequate warning systems	6. Fire or explosion hazards
7. Poor housekeeping; disorder	8. Hazardous environmental conditions: gases, dusts, smokes, fumes, vapours	9. Noise exposure

(*Continued*)

Table 3.1 (*Continued*)

10. Radiation exposure	11. Temperature extremes	12. Inadequate or excess illumination
13. Inadequate ventilation	14. Inadequate instructions/procedures	15. Other unsafe condition

Personal Factors

1. Inadequate physical/physiological capability	2. Inadequate mental/psychological capability	3. Physical or physiological stress
4. Mental or psychological stress	5. Lack of knowledge	6. Lack of skill
7. Improper motivation		

Control Failures

1. Failure to prevent the initial build-up of the energy or substance (Elimination)	2. Failure to reduce the amount of energy or substance stored or build-up (Substitution)	3. Failure to prevent the release of the energy or substance (Engineering or administrative control)
4. Failure to reduce the rate of spatial distribution of release of energy or substance from its source (Engineering control)	5. Failure to separate in space or time the energy or substance being released from the susceptible structure or person (Engineering or administrative control)	6. Failure to separate the energy being released from the susceptible structure or person by interposition of an effective barrier (Engineering control)
7. Failure to decrease the hazardous nature of contact surface, subsurface, or basic structure which can come into contact with susceptible structure or person (Substitution)	8. To strengthen the structure or person which might be damaged by the energy transfer or contact with substance (Personal protective equipment)	9. To move rapidly in detection and evaluation of damage and to counter its continuation and extension (Administrative control and emergency preparedness)
10. All those measures which fall between the emergency period following the	11. Failure to ensure that relevant employees have sufficient awareness, knowledge,	12. Failure to provide adequate supervision (Administrative control)

(*Continued*)

Table 3.1   (*Continued*)

damaging energy exchange and the final stabilization of the process (Emergency preparedness)	and/or skills (Administrative control)	
13. Other control failures		

Underlying Factors		
Leadership (the buck stops here)		

1. Failure to take accountability	2. Failure to maintain oversight of SHE* management system	3. Failure to maintain oversight of SHE* culture
4. Lack of SHE* knowledge	5. Lack of SHE* leadership skills	6. Lack of commitment or inappropriate attitude or motivation towards SHE
7. Mental or psychological stress	8. Failure to demonstrate commitment to SHE*	9. Other leadership factors

Management System Inadequacies (based on OHSAS 18001:2007/ ISO 45001 and ISO 14001:2015)		
1. Policy	2. Structure, roles, responsibilities and authorities	3. Risk assessment and identification of controls
4. Processes to establish objectives and programmes	5. Resource allocation processes	6. Processes to ensure competence and awareness
7. Communication and consultation processes	8. Documentation processes and requirements	9. Processes for planning, coordination and control of operations
10. Processes to ensure emergency preparedness	11. Processes for monitoring, measurement, analysis and evaluation of system performance	12. Processes to manage incident, non-conformity, corrective action and preventive action
13. Other management system inadequacies		

(*Continued*)

Table 3.1 *(Continued)*

SHE* Culture (Shared SHE* values and beliefs in the organisation or at different organisational levels)		
1. SHE* is given lower priority as compared to other organizational goals, e.g. productivity, deadlines and profit	2. Believes that incidents will not happen or insensitivity to SHE* hazards	3. Unsafe behavioural norms or practices
4. Other SHE* culture issues		

*SHE = Safety, Health and Environmental [management]

analyses. The technique is an extension of incident causation models like the Energy Transfer Model (ETM), the Loss Causation Model (LCM) and the Swiss Cheese Model, but unlike these models, the ECT is meant to be a structured, simple and flexible step-by-step incident analysis technique. Investigators and managers using the ECT should be able to determine a comprehensive set of causes and factors so as to recommend actions to prevent recurrence of an incident.

The ECT has five main components: (1) work context, (2) incident sequence, (3) direct causes, (4) control failures and (5) underlying factors. The following sub-sections will explain each of these components in more detail.

3.5.1 Work context

Work context describes the workplace scenario where the incident took place. Descriptors of work context can include the type of work being executed, the types of trades and workers involved, the equipment and material used in the work, the work environment where the work took place and other interacting or nearby work. The descriptors can be used to categorise incident cases during incident data analysis. The work context, or parts of the work context, together with the direct causes are the prerequisites of an incident.

3.5.2 Incident sequence

Incident sequence is a concept adapted from the Energy Transfer Model (ETM) described in the earlier chapter. It splits the incident into four types of key events: breakdown event, intermediate event, contact event and consequence. The types of losses and events in the Loss Causation Model (LCM) are useful references for the incident sequence. The incident sequence describes what happened from the initiating point of the accident till the occurrence of the consequences, so that opportunities for preventing future occurrence of the incident can be identified.

A breakdown event is defined as an initiating point of loss of control of a source of energy or substance that, without an intervening event (e.g. presence of a control), will lead to the occurrence of a contact event. In contrast, a contact event is an event where the victim comes into contact with the source of hazardous energy or substance. An intermediate event refers to any significant event between the breakdown event and contact event that if adequately controlled will prevent or mitigate the consequences of the incident. It is not necessary to include an intermediate event (i.e. it is optional), or there could be more than one intermediate event if the incident is complex. The key criteria for deciding if intermediate events should be captured in the incident sequence is if there are opportunities for controlling the intermediate event. If the intermediate event represents an opportunity to put in a control to prevent future occurrence of the incident, then it must be described in the incident sequence. On the other hand, an intermediate event may be included to ensure that the incident sequence is complete and logical to the reader, but it may not provide an opportunity for control. Consequences refer to the undesirable effects of the incident, e.g., property loss, environmental pollution, number of man-days lost and type of injury.

It is beneficial to define incidents based on the incident sequence structure, so that causal factors and controls can be classified systematically into three levels of intervention and causation, namely, "post-contact", "pre-contact" and "prevention" levels. By focusing on the three levels of intervention and causation, investigators will be encouraged to broaden their scope of investigation to identify more opportunities for improvement. For example, when a worker loses his or her balance and falls off the edge of a building, an investigator could easily state that the 'main cause' of the accident is that the worker was not using the fall arrest system provided. Even if the underlying factors and the controls that had contributed to the contact event (striking the ground) were identified, the recommendations based on the investigation would probably only prevent the recurrence of the contact event, but not the breakdown event (i.e. the worker's loss of balance, in this instance). To effectively lower the risk of the activity, the factors that contributed to the occurrence of the breakdown event, the intermediate event, the contact event and the consequences of the incident should all be identified.

The event just prior to the breakdown event, i.e. the preceding event, can be included to better describe the incident sequence. The preceding event would help to clarify the work context.

3.5.3 Direct causes

As in the case of immediate causes of the Loss Causation Model (LCM), direct causes refer to causes that are directly linked to an event in the incident sequence. A direct cause is a necessary condition for an event to occur. When the cause is removed, the event will not occur. Direct causes are classified into unsafe conditions, unsafe acts and personal factors. Unsafe conditions refer to any direct cause related to non-human aspects, e.g., faulty equipment, unsuitable material and unsafe work environment. Unsafe acts refer to observable actions or lack of action of front-line personnel executing the work. In most instances, the definitions of safe and unsafe acts are dependent on risk assessments and relevant standards and legislations. In broad terms, a condition or action that leads to an unacceptable risk level is unsafe. When standards and legislations are applicable, non-compliance is usually assumed to define what is unsafe.

Personal factors refer to factors such as the physiological, mental and psychological factors of a person (Bird *et al.*, 2003). Some examples include inadequate strength, insufficient physical endurance, lack of knowledge and improper motivation. Within the context of direct causes, it is the personal factors of front-line workers or people directly involved in the incident that are considered. For example, consider a worker who performed an unsafe act by climbing up the bracings of an access scaffold. A common personal factor that led to this unsafe act is improper motivation to save time and effort.

It should be noted that direct causes may not be present if the cause of the incident event is a control failure. This means that the incident event directly links to the control failure.

3.5.4 Control failures

Control failures refer to failures of specific controls that an organisation has or should have. These controls should have been highlighted in a risk assessment that the organisation conducted. Failure occurs when there is a lack of control, i.e. control was not identified in the relevant risk assessment), inadequate control, i.e. control was identified but it was not adequate in preventing the direct cause and/or incident event, and inadequate execution of control, i.e. the control could have prevented the direct cause and/or incident event, but it was not implemented. The possible control failures in Table 3.2, which were extracted from Table 3.1, are based on the ten counter-measures of the Energy Transfer Model (Haddon Jr, 1973). The relevant incident events, i.e. breakdown event, intermediate event, contact event, and consequence, and corresponding level of intervention and causation were also identified in the table. The relevant ECT events refer to the incident event that can occur if the control failure occurs. However, with a wide variety of possible incident sequence and control failures, Table 3.2 is only a guide and exceptions are expected in some cases.

Some examples of controls are edge barricades, warning signs, training, personal protective equipment and emergency evacuation paths.

Table 3.2 Types of control failures

Possible Control Failures	Relevant ECT Event (Typical)	ECT Level of Intervention & Causation
1. Failure to prevent the initial build-up of the energy or substance	BE	Prevention
2. Failure to reduce the amount of energy or substance stored or built-up	BE, CSQ	Prevention, post-contact
3. Failure to prevent the release of the energy or substance	BE, IE	Prevention
4. Failure to reduce the rate of spatial distribution of release of energy or substance from its source	IE, CE	Pre-contact
5. Failure to separate in space or time the energy or substance being released from the susceptible structure or person	IE, CE	Pre-contact
6. Failure to separate the energy being released from the susceptible structure or person by interposition of an effective barrier	IE, CE	Pre-contact
7. Failure to decrease the hazardous nature of contact surface, subsurface, or basic structure which can come into contact with susceptible structure or person	IE, CE	Pre-contact
8. To strengthen the structure or person which might be damaged by the energy transfer or contact with substance	IE, CE, CSQ	Pre-contact, post-contact
9. To move rapidly in detection and evaluation of damage and to counter its continuation and extension	CE, CSQ	Post-contact
10. All those measures which fall between the emergency period following the damaging energy exchange and the final stabilization of the process	CSQ	Post-contact
11. To ensure relevant employees have sufficient awareness, knowledge, and/or skills	BE, IE, CE, CSQ	Prevention, pre-contact, post-contact

BE — Breakdown event; IE — Intermediate event; CE — Contact event; CSQ — Consequence

Controls are usually identified during a risk assessment or risk management process, which will be covered in Chapter 4. Controls need to be specific and must have direct relation to the direct cause and/or incident event. A lack of control refers to the absence of a control, for example, the absence of necessary barricades. Inadequate control refers to the presence of a control, but the design or planning of the control was not adequate. For example, a barricade was designed with insufficient height and the open edge becomes inadequately protected, hence an unsafe condition. Inadequate execution of control refers to non-compliance with the initial plan or design of a control. The unsafe act of using a non-explosion-proof torchlight in an area designated as a high fire or explosion risk zone can be due to an inadequately executed briefing (the control). In this example, the briefing could have failed to cover some of the required content, including the locations of the high fire-risk zones.

3.5.5 Underlying factors

Underlying factors refer to organisational and managerial factors that are contributory to the failure of controls and occurrence of direct causes. They are similar to the latent conditions in the Swiss Cheese Model, and basic causes and lack of control in the Loss Causation Model (LCM). Underlying factors are often contributory in nature and their determination may have to depend on investigators' professional judgment, but identifying the underlying factors can lead to a broader and more sustainable impact on safety and health performance. Three types of underlying factors are identified in Figure 3.2: leadership, management system and culture.

In the context of WSH management, leadership refers to the attributes of senior managers that have an influence on WSH. A senior manager is a person who has the ultimate authority and accountability, i.e. the buck stops with him or her. The leadership issues highlighted in Table 3.1 are possible factors that can contribute to the control failures and direct causes. Leadership issues will be discussed in more detail in the Chapter on Safety culture.

A management system refers to a set of policies, processes and structures that interact to assure or support Safety, Health and Environmental [management] (SHE) controls so that they maintain or improve the SHE performance of the organisation. In Table 3.1, we assumed that the breakdown event in an accident was the worker tripping over a wooden plank (loss of balance). The breakdown event was due to poor housekeeping (an unsafe condition) and the control failure is the failure to conduct frequent housekeeping in the worksite (i.e. failure to prevent build-up of hazardous "substance"). A relevant management system inadequacy could be inadequate processes for monitoring, measurement, analysis and evaluation of the system's performance. More specifically, although the site might have defined housekeeping schedules, there was a lack of inspections to check that housekeeping was actually conducted. Management system inadequacies are usually inadequacies in the plan–do–check–act cycle.

Another important underlying factor is organisational culture. Safety culture has gained a lot of attention in the WSH literature over the years. Safety culture can be seen as a subset of organisational culture, and the latter is defined as, "a system of shared values [what is important] and beliefs [how things work] that interact with a company's people, organisational structures, and control systems to produce behavioural norms [the way we do things around here]" (Uttal 1983). When applied to safety and health management, the "shared values", "[shared] beliefs" and "behavioural norms" refer to safety-related values, beliefs and behavioural norms (see Table 3.1).

Organisational culture is very abstract and even people within the organisation may not be able to explain their own culture very well, but they simply know that the culture is there. During the Safety Management in Context conference in 2013, Edgar Schein, who is an organisational culture guru, recommended that safety practitioners should focus on [management] processes and not 'safety culture'. He was also sceptical about the concept of safety culture, which he felt was an ill-defined concept. If we take Schein's comments at face-value, it seems to imply that the focus of any investigation of underlying factors should be on management processes. However, it is believed that Schein's comments were not meant to dilute the importance of organisational culture in incident prevention. His comments highlighted the importance of focusing on management systems so that shared values, beliefs and norms can be better managed. The reality is that the concept of safety culture is abstract and its assessment is usually subjective. Furthermore, culture constantly interacts with organisational structure processes, i.e., the organisation's management system. A management system is a reflection of the culture and vice versa. Thus, instead of focusing on the abstract, it is better to focus on the concrete. Nevertheless, it is recommended that when applying the ECT, if there is sufficient information to infer the shared beliefs, values and norms that contribute to an incident, the cultural problems and issues should be highlighted. Safety culture will be discussed in more detail in Chapter 7.

3.5.6 Implementing the Event Causation Technique (ECT)

The above described the ECT framework. To utilise the ECT, a series of steps should be conducted systematically during the analysis stage of the investigation. The steps in the application of ECT are as follows:

1. Describe the work context in which the incident occurred.
2. Create a timeline of the incident. The initial timeline should be as comprehensive as possible.
3. Select the key events (i.e. breakdown event, intermediate event(s), contact event and consequences) for detailed analysis.
4. For each of the key events in the incident sequence, a series of 'whys' are asked to facilitate the identification of direct causes, control failures and underlying factors. In accordance with the principle of multi-causation, within each component of the ECT, there can be more than one answer to the 'why'. The asking of 'why' questions should only stop when the answers reach underlying factors, i.e. leadership, management system, and/or culture.
5. For communication and presentation purposes, it is necessary to simplify the why-analysis into the ECT diagram (see Figure 3.2 and Figure 3.5) to ensure that readers can understand the incident causation

in one diagram. This involves compressing the why-analysis into more generic words and phrases (based on Table 3.1).

Table 3.1 is a useful prompt for applying the ECT, but during the why-analysis and creation of the ECT diagram, the description of causes and factors should be based on the generic categories in the table. Each cause and factor should be clear enough to help readers understand what you are referring to. For example, if a relevant unsafe act is "failure to secure", the generic category should be followed a brief description such as, "did not secure dump truck tailgate".

3.5.7 Concrete hopper accident

The following case study, Case Study 2, demonstrates how the ECT can be applied in an accident investigation.

3.5.7.1 *Describe the work context*

A worker was working inside a man-cage attached to a concrete hopper. The concrete hopper is a container used to carry wet concrete and it has a release lever to allow workers to pour the concrete into formworks or other containers. In this case, the worker is needed to pull the lever to pour the concrete into a joint between two precast wall panels. The concrete hopper was lifted by a tower crane and the man-cage was connected to the concrete hopper. The work context can be summarised into the following statement, "Worker pouring concrete into precast joint while standing in man-cage lifted by tower crane and using concrete hopper". The statement points out the type of activity that was being conducted, and the equipment and plant that were used.

3.5.7.2 *Create timeline*

The timeline in Table 3.3 shows the key events related to the case. The focus was very much on the day of the accident, but relevant information about the date of purchase and history of the hopper and man-cage, the

Table 3.3   Timeline for Case Study 2

Date	Time	Event
1 May 2012		Hopper first installed with man-cage
May–Sep 2012	—	Man-cage used on several joint castings
2 September 2012	0830	Worker starts work in man-cage; crane operator lifts man-cage
	0832	Hopper was lifted to the 6–7th level joint
	0835	Crane operator needs to adjust the height of the hopper to facilitate concreting
	0836	Hopper became dislodged suddenly
	0840	Supervisor contacted WSHO
	0841	WSHO called ambulance
	0850	Worker sent to hospital

start date of the project and other relevant activities, and the date when the operator and relevant workers first began working on site can also be included.

3.5.7.3 *Describe incident sequence*

The incident sequence is created based on the timeline, which was based on a range of evidence. The events that are uncertain or were based on the opinions of investigators need to be highlighted using a symbol or a different colour. These uncertainties can then be further validated using new evidence. The incident sequence shown in Figure 3.3 is the finalised incident sequence.

3.5.7.4 *Conduct why-analysis*

Once the incident sequence is established, the why-analysis should be conducted on each incident event. Though in Case Study 2, the analysis was focused on the breakdown event — and investigators can choose to focus on some of the key events due to time constraints or assessment of available evidence — **it is emphasised that each incident**

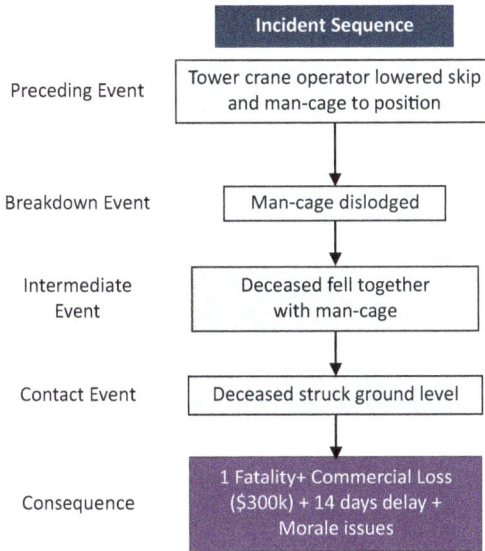

Figure 3.3   Case Study 2 incident sequence

event should be analysed as much as possible. With reference to Figure 3.4, the qualitative analysis process is based on one simple question, "why did the event or cause occur?" For instance, "why did the man-cage dislodge from the concrete hopper?" The man-cage dislodged because when the crane operator was lowering the hopper and man-cage into position (preceding event), the man-cage unhooked from the hopper (the cause for the breakdown event).

Two basic guiding principles need to be consulted for each cause, firstly, whether the cause is "necessary". The first guiding principle helps to check whether the cause or factor is a true cause. If it is not "necessary" (i.e. if it does not influence the existence of the event), then it should be removed. The second checks whether the cause is "sufficient". A cause may be necessary, but it is not sufficient to cause the event, in which case another cause needs to be included.

Coming back to Case Study 2, during the activity of lowering the hopper, the unhooking of the man-cage is necessary to cause the man-cage to be dislodged. This means that without the unhooking action,

Work Context

Worker pouring concrete into precast joint while standing in man-cage lifted by tower crane and using concrete hopper

Incident Sequence

Preceding Event: Tower crane operator lowered hopper & man-cage to position

Breakdown Event: Man-cage dislodged

Intermediate Event: Deceased fell together with man-cage

Contact Event: Deceased struck ground level

Consequences: 1 Fatality + Commercial Loss (S$300k) + 14 days delay + Morale issues

— Why? — Man-cage sat onto precast wall
— Why? — Signalman error
— Why? — Lack of supervision

— Why? — Man-cage unhooked from hopper
— Why? — Fatigue
No working hours policy

Lifting procedures for man-cage (high risk lifting) inadequate

— Why? — Unsecured connection
— Why? — No checks by Authorised Examiner (AE)
— Why? — Checks by AE are only implemented for some Lifting Equipment (LE) but not for all LEs including this man-cage (Inadequate implementation)

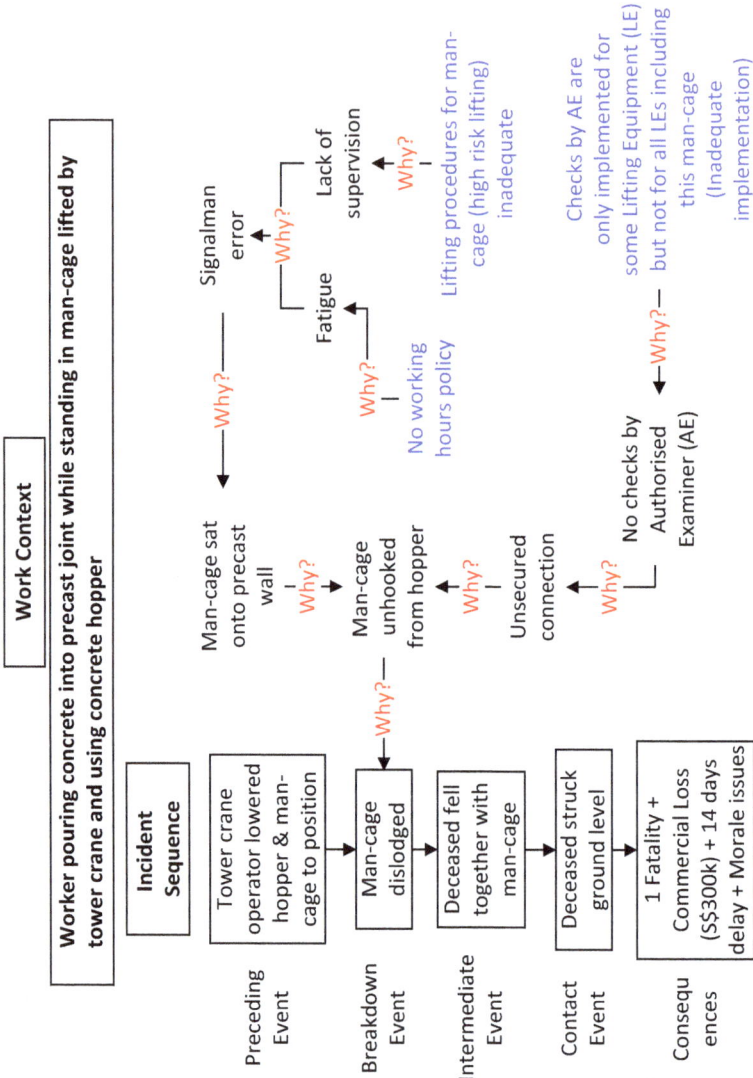

Figure 3.4 Why-analysis for the breakdown event

the man-cage will not be dislodged. The unhooking is also sufficient to cause the dislodgement, i.e. other causes like strong wind or high temperature do not need to be present to result in the man-cage dislodging. The interrogation continues for "man-cage unhooked from hopper". In this case, two causes were identified. Firstly, in the midst of the lowering process, the man-cage sat onto the precast wall, which produced a vertical reaction force, contributing to the man-cage becoming unhooked. Thus, "Man-cage sat onto precast wall" is a necessary cause. However, for the man-cage to 'complete' the unhooking action, pushing the man-cage against the precast wall alone is not sufficient to produce the "unhooking". The connection between the man-cage and hopper must have been unsecured, which allows a vertical force to push the man-cage out of the hopper. Thus, "unsecured connection" is included as another cause. The same process continues for each cause, guiding the investigator in his or her evidence-gathering process. The why-analysis can be guided by the taxonomy in Table 3.1, but the investigators should not be constrained by the taxonomy. The why-analysis should end when the investigator uncovers the underlying factors that can produce fundamental systemic improvement to the organisation.

3.5.7.5 *Summarise using ECT diagram*

The why-analysis can extend into several pages and can be confusing to the readers, especially personnel not directly involved in the accident. Thus, the why-analysis is summarised into a simpler ECT diagram (Figure 3.5). The ECT diagram should also include the summary of the analysis for the contact event and consequences. This is when each column in the ECT diagram, incident sequence, direct causes, control failures and underlying factors, are filled in based on the why-analysis. Examples of complete ECT diagrams can be found in Goh and Soon (2014).

Using the ECT diagram, recommendations can then be made to improve the SHE performance of the organisation. Each recommendation

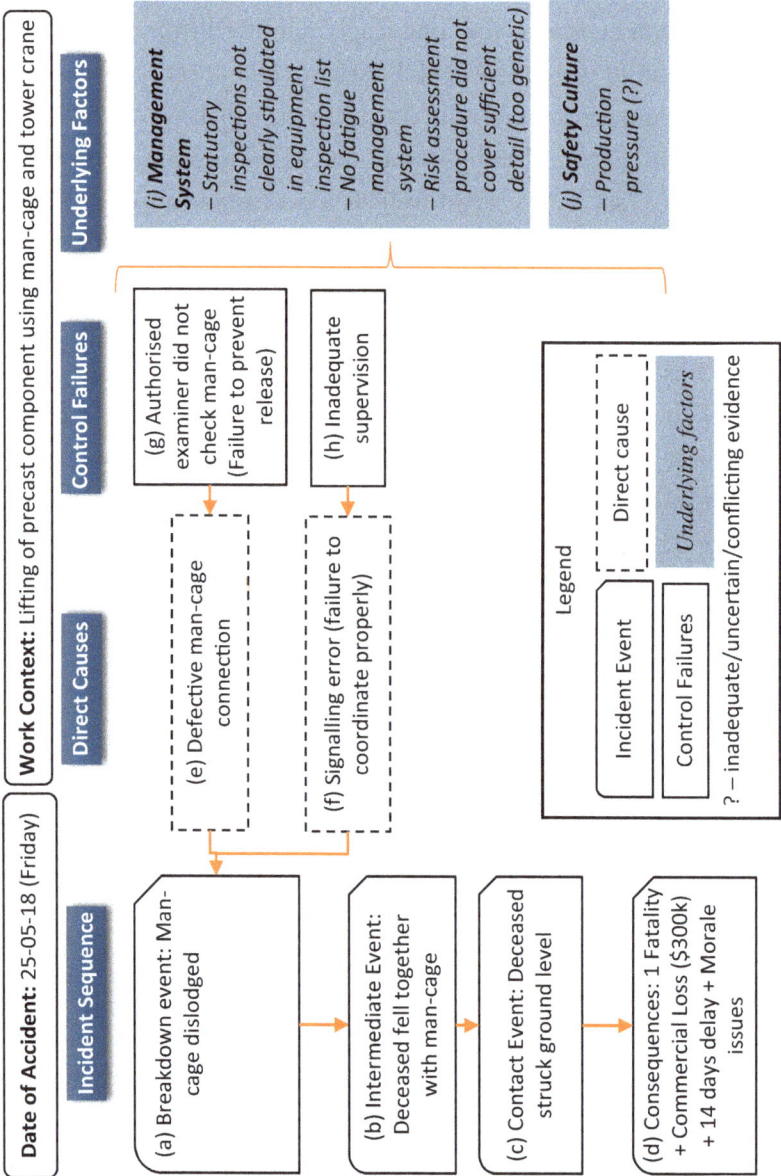

Figure 3.5 ECT diagram summarising the information in why-analysis

should be linked to different parts of the ECT diagram. Based on the ECT diagram in this example, some possible recommendations are:

1. Top management to impose an operational requirement for engineers and supervisors to ensure that statutory equipment are checked before they are used.
2. Human resource to develop a set of work hour policies and a fatigue management system that is applicable to all site personnel (including sub-contractors); policies to be endorsed by top management.
3. To review lifting procedures and risk assessment to differentiate between low- and high-risk lifting operations.
4. To review risk assessment (RA) procedures to determine when generic RA is allowed.
5. To ban all man-cage usage until risk assessment for lifting is reviewed; sites to use scaffold as substitution in the interim.

The clear linkage provides a stronger rationale for the recommendation. The recommendations can then be monitored using an action plan that clearly answer the following:

- Who to do what and by when?
- When is the next review date?
- Who to monitor progress?
- Who is the management representative to ensure progress?
- How to track effectiveness of measures?

More examples of ECT diagrams can be found in Chapter 10.

3.6 Summary

Incident investigation is an important part of workplace safety and health (WSH) management. When an incident happens, there must be deliberate attempts to identify the patterns in the way things were done so that these patterns can be changed to prevent a recurrence of the incident. These patterns occur because of some underlying factors such as the way the management system is constructed and the characteristics of the safety culture of the organisation. Thinking of how the system influenced the occurrence of an event is a key concept in systems thinking and it is an important skill for any manager.

The investigation process is a qualitative, systematic and evidence-based process. The investigator needs to connect the investigation, a reactive process, with proactive risk assessment so that relevant risk controls identified in the risk assessments are reviewed and future risk assessments are strengthened. The event causation technique (ECT) is a simple and flexible method for the incident analysis. Despite its simplicity, it is meant to be thorough and the final ECT diagram should provide a useful summary of an incident.

Review Questions

1. "An accident is a matter of luck." Discuss the validity of this statement and how management should see the role of luck in accidents.
2. Explain the difference between corrective actions, preventive actions and actions for continual improvement.
3. Explain how incident investigation and risk assessment are related.
4. Explain what it means for investigators to understand the "retrospective" nature of incident investigation.
5. Provide examples of the four types of evidence in an incident investigation.
6. Practise the ECT method on an accident case that you find on the internet.

References

Bird, F. E., Germain, G. L., and Clark, M. D. (2003). *Practical loss control leadership*. Duluth, Georgia: Det Norske Veritas (U.S.A.), Inc.

British Standards Institute (2007). "BS OHSAS 18001:2007 Occupational health and safety management systems — Requirements." BSI, London.

British Standards Institution (2004). "BS EN ISO 14001:2004 Environmental management systems — Requirements with guidance for use." British Standards Institution, London.

Chua, D. K. H., and Goh, Y. M. (2004). "Incident causation model for improving feedback of safety knowledge." *J. Constr. Eng. and Manage. — Am. Soc. of Civ. Eng.*, 130(4), 542–551.

Dekker, Sidney. *The field guide to understanding 'human error'*. CRC Press, 2017.

Energy Institute (2008). "Guidance on investigating and analysing human and organisational factors aspects of incidents and accidents." <http://www.energyinstpubs.org.uk/tfiles/1368180505/817.pdf>. (May 10, 2013).

Goh, Y. M., and Soon, W. T. (2014). *Safety Management Lessons from Major Accident Inquiries*, Pearson, Singapore.

Haddon Jr, W. (1973). "Energy damage and the ten countermeasure strategies." *Human Factors*, 15(4), 355–366.

Uttal, B. (1983). "The corporate culture vultures." *Fortune*, Oct. 17, 66–72.

Chapter 4
Workplace Safety and Health Risk Management

The Event Causation Technique

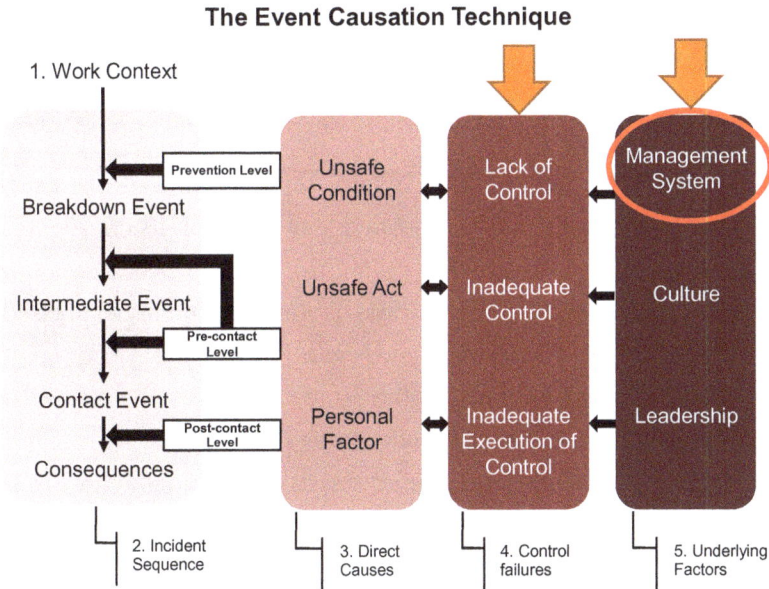

Risk management (RM) is an important aspect of the management system as it forms the planning part of the workplace safety and health (WSH) management framework (see Event Causation Technique (ECT) diagram above). The end products of an RM process are the controls that are supposed to prevent incidents.

4.1 Introduction

Risk management is not unique to workplace safety and health (WSH) management. According to the ISO standard, BS ISO 31000:2009

Risk Management — Principles and Guidelines (British Standards Institute 2009), **risk is the effect of uncertainty on objectives**. BS ISO 31000:2009 also provided the following definitions and comments:

- An effect is a positive or negative deviation from the expected.
- Objectives can refer to different aspects of an organisation such as financial, productivity, workplace safety and health (WSH), and environmental aspects. At the same time, objectives can be at different levels of an organisation, for example, the strategic, project, product and process levels.
- Uncertainty is the partial or complete deficiency of information related to understanding or knowledge of an event, its consequences or likelihood.
- In accordance to its definition, risk is most commonly expressed as a combination of the consequences of an event (its severity, costs or benefits) and the likelihood of the event.
- Risk management is a systematic process of identifying, analysing, evaluating, controlling and monitoring risk. It is implemented to help assure achievement of organisational objectives.

In the context of this book, the objectives we are concerned with are WSH objectives. When organisations assess WSH risk, they typically refer to negative effects such as accidents and pollutions, but there are also potential positive effects, like higher employee morale, higher productivity, and improved reputation. However, these positive effects are only apparent at organisational or strategic levels. Most of the WSH risk management is conducted at the operational level and focuses on operational issues. We will first discuss the importance of risk management in WSH and the role that it plays. We will then cover the risk management process and key WSH hazards and controls. Finally, we will discuss some of the problems and difficulties with WSH risk management.

4.2 Workplace Safety and Health Risk Management and Assessment

Risk management, which includes risk assessment, is a cornerstone for the performance-based approach to the regulation of workplace safety and health (WSH). Risk management is sometimes used inter-changeably with risk assessment, but there are differences that will become obvious when we discuss the steps in risk management subsequently.

4.2.1 Background

One of the key legislations that resulted in the common use of risk assessment in WSH is the UK Roben's report in 1972, which highlighted the following:

- Health, safety and welfare at work could not be ensured by an ever-expanding body of legal regulations enforced by an ever-increasing army of inspectors;
- The primary responsibility for ensuring health and safety should lie with those who create risks and those who work with them; and
- The law should provide a statement of principles and definitions of duties of general application, with regulations setting more specific goals and standards.

Singapore adopted a similar approach in 2006 when the WSH Act was enacted. At the time of the enactment, Minister of Manpower Dr Ng Eng Hen highlighted the following principles:

- Reduce risks at source
- Promote industry ownership of standards and outcomes
- Penalise poor management

Thus, for the industry to reduce risk at source and own standards and outcomes, employers and duty holders are expected to conduct risk

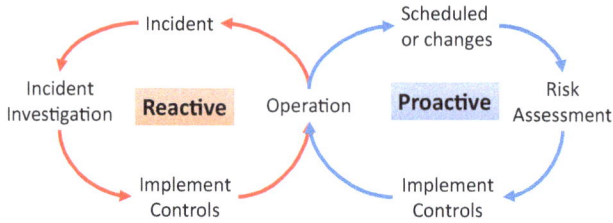

Figure 4.1   Risk assessment allows stakeholders to be proactive

assessment (as part of their risk management) to proactively reduce the WSH risk of its operations and activities (see Figure 4.1). Even though incident investigation is an important aspect of WSH management, it is dependent on the occurrence of incidents which could have resulted in severe injuries and losses. Thus, risk assessment, instead of incident investigation, should always be the key focus of WSH management. Risk assessments must be scheduled prior to all operations so that controls can be implemented. Risk assessments should also be conducted whenever there are changes in the operations, including after significant incidents.

4.2.2 Risk assessment and incident causation models

The risk assessment process is very much related to incident causation models. This is because incident causation models tell us why incidents happen and risk assessment aims to identify possible incidents, their causes and corresponding controls and systems to manage the likelihood and/or severity of the incidents (see Figure 4.2). Thus, during risk assessment we should put in controls to manage the basic causes and immediate causes (hazards) to prevent the incidents.

The possible controls can be guided by Haddon's Energy Transfer Model (ETM; see Chapter 2) and the hierarchy of control, which will be discussed later.

4.2.3 Conceptual basis of risk assessment

The UK Health and Safety Executive (HSE) published the seminal document, "Reducing Risk, Protecting People" in 2001. It provides a

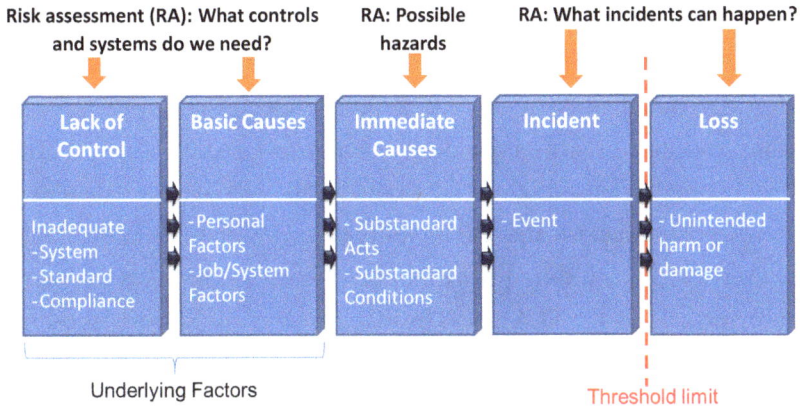

Figure 4.2   Risk assessment and incident causation model

useful conceptual basis for WSH risk assessment. Some key concepts presented include the following:

- Risk is a given — there are hazards everywhere, even if we do not do anything. Thus, it is theoretically impossible to have zero risk.
- Factors influencing risk perception of man-made hazards:
  - if a hazard poses a risk to things that are important to us, we will perceive the hazard to be higher risk.
  - how well the process (giving rise to the hazard) is understood (knowledge and competency)
  - how equitably the danger is distributed ("are we the only ones that are exposed to the risks?")
  - how well individuals can control their exposure ("do we have the resources and capabilities to control the hazards?")
  - whether risk is assumed voluntarily ("did we choose to be exposed to this risk or was it imposed on us?")

As seen, risk assessment is heavily influenced by human minds, cultures, judgement and values. Thus, it is difficult to establish the "objective and true risk", which may not even exist. In the end, it is about tolerability and acceptability of risks by individuals, organisations and society. In addition, a major factor is the amount of resources (time, money, etc.) that should be invested to control the hazards, as compared to the aggregated consequences suffered by society, community, organisations and groups, when accident(s) and/or ill health occur(s). Thus, risk assessment is essentially a cost-benefit analysis, which balances the costs of controlling risk and the benefits of having the risk controlled.

Another key guiding principle of risk assessment is "as low as reasonably practicable" (ALARP). The risk levels of all hazards should be ALARP, which means that the costs involved in reducing the risk level further would be grossly disproportionate to the benefits gained. With reference to Figure 4.3, the risk level of a hazard can be placed on a scale with *Broadly acceptable*, *Tolerable* and *Unacceptable* regions. The *Unacceptable* region is where the risk level of a hazard is too high and no person should be exposed to this risk regardless of the benefits gained. On the other hand, the *Broadly Acceptable* region indicates that the risk level is widely regarded by society to be acceptable because the inherent risk is low, or the controls are very effective. The *Tolerable* region is where the risk level is significant, but tolerable such that the benefits justify the exposure to the hazard. Some examples of benefits include employment, lower cost of production, personal convenience, and continued supply of basic needs like energy, food and water. In each of these regions, the risk level of the hazard should be kept ALARP. The concept of ALARP is especially important in the *Tolerable* region. For a hazard to be tolerated, it must be properly risk assessed, the residual risk level must be ALARP, and the risk level must be periodically reviewed to ensure that it is ALARP.

4.3 Risk Management Regulations

In Singapore, all employers are required to comply with the WSH (Risk Management) Regulations 2007 (RM Regulations). The RM Regulations is a very concise legislation containing only eight regulations, and is a very fundamental regulation in the WSH regime. Anyone that employs

Figure 4.3   HSE Framework for Tolerability of Risk; adapted from http://www.hse.gov.uk/risk/theory/r2p2.pdf (Pg 42)

or engages others for work has to conduct risk assessment (see RM Regulations). Risk assessment is defined as "the process of evaluating the probability and consequences of injury or illness arising from exposure to an identified hazard, and determining the appropriate measures for risk control". A hazard means anything with the potential to cause bodily injury, and includes any physical, chemical, biological, mechanical, electrical or ergonomic hazard. According to the RM Regulations, risk means the likelihood that a hazard will cause a specific bodily injury to any person. This definition is more specific than that provided by BS ISO 31000 because it is contextualised to WSH.

Employers, self-employed and principals[1] are required to "eliminate foreseeable risk", but if it is not "reasonably practicable" to eliminate the risk, employer, self-employed person or principal are to implement

---

[1] In accordance to the Workplace Safety and Health Act, "principal" means a person who, in connection with any trade, business, profession or undertaking carried on by him, engages any other person otherwise than under a contract of service (i.e. an employee) — to supply any labour for gain or reward or to do any work for gain or reward. Typically, a principal refers to the client in a client-contractor relationship.

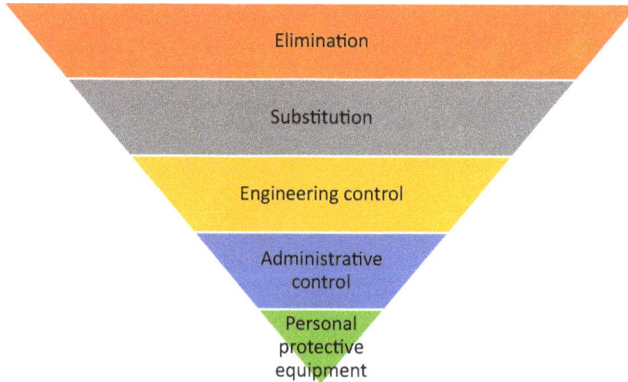

Figure 4.4   Hierarchy of control

"reasonably practicable measures to minimise the risk" and safe work procedures to control the risk. The control measures identified in the RM Regulations can be found in Figure 4.4, which is also known as the hierarchy of control. The control measures are defined as follows:

- "substitution" means the replacement of any hazardous material, process, operation, equipment or device with less hazardous ones.
- "engineering control" means the application of any scientific principle for the control of any workplace hazard; and includes the application of physical means or measures to any work process, equipment or the work environment, such as the installation of any barrier, enclosure, guarding, interlock or ventilation system.
- "administrative control" means the implementation of any administrative requirement which includes a permit-to-work system.
- Personal protective equipment (PPE) refers to equipment such as helmets, safety boots, harnesses and safety goggles.

As noted earlier, Haddon's energy transfer model is also a useful guide for the identification of possible risk controls.

The hierarchy of control indicates that elimination is the most effective approach, while personal protective equipment (PPE) is the least effective. This is because PPE only minimises the impact of hazards and is heavily reliant on the user using the PPE correctly at all times. In contrast, elimination removes the hazard and the need for any form of control. Substitution and engineering control are highly desirable, but many-a-times, organisations rely heavily on administrative controls and PPE to manage WSH risk because they are more easily implemented. It should be noted that the control measures should be implemented concurrently to reduce the consequences if any of the control measures fail unexpectedly. For example, a worker working at height should be protected by barricades, but if the worker needs to reach over the barricade during the course of his/her work, then a fall protection system (full body harness, lanyard with personal energy absorber and anchorage), which is a form of PPE, should be used by the worker to reduce the risk of falling from height.

Table 4.1 shows some examples for each type of control in the hierarchy of control.

"Safe work procedure" (SWP) is frequently seen as a type of administrative control measure but the Technical Advisory for Working at Height (Workplace Safety and Health Council 2008) describes SWP as,

> "a set of systematic instructions on how work can be carried out safely. Arising from the risk assessment, a set of SWP should be written for various jobs on site. The SWP provides a step by step account of how jobs are to be executed, who is in charge of these jobs, what safety precautions must be taken (based on the risk assessment made earlier) and what kind of training is necessary for the workers doing these jobs. The permit-to-work system has to be integrated with the Safe Work Procedure so that the supervisors are made aware of the safety requirements and checks. The SWP must be communicated to everyone involved in the job so that each is aware of the role they play in it."

Therefore, an SWP is not a control measure, but a documentation of different control measures. Thus, for it to be effective, it must be written in a very detailed fashion and must include the risk controls to be taken in the course of the work and during an emergency. In addition, SWPs must

Table 4.1   Examples of controls

Type of Control	Example
Elimination	• Using long arm roller paint brush instead of working on scaffolding • Designing windows that can be cleaned from indoors (e.g. can be turned inwards) to eliminate the need for workers to clean the windows from outside and at heights
Substitution	• Substituting flammable and potentially toxic solvent-based paints (i.e. may release flammable Volatile Organic Compounds) with water-based (also known as acrylic emulsions) paints • Substituting a noisy machine (>85 dBA) with a less noisy machine (~60 dBA)
Engineering control	• Barricades along open edge • Ventilation to remove toxic chemicals in the atmosphere • Sensors to detect human motion in the vicinity of a machine and stopping the machine when the sensors are triggered
Administrative control	• Work at heights training for workers working in mobile elevated work platforms • Permit to work for workers entering confined space • Warning signs at the bottom of a scaffold under erection to remind workers of the need to stay away
Personal protective equipment (PPE)	• Use of ear plugs by workers working in noisy area(s) • Use of eye protection for workers using cutting machines • Dust mask for workers cleaning dusty areas

highlight suitable personal protective equipment that must be provided to the persons carrying out the work. Thus, the SWP is essentially a set of work instructions that includes risk controls identified in the risk assessment. There is no fixed format for SWPs, but they should be written such that they are easily comprehended by the people conducting the work — like how assembly instructions for self-assembled furniture include step-by-step guidance on how to put the furniture together as well as safety measures (such as providing a mat to protect the person assembling the furniture from injuries such as knee injuries due to prolonged kneeling on a hard floor).

Risk assessment (RA) is expected to be recorded, disseminated and reviewed. The documentation of RA is necessary because it helps to ensure a detailed analysis of the work and establishes accountability, but there are many challenges in actual implementation, which will be discussed subsequently. RA is legally required to be reviewed every three years, when there is an incident, and when there is significant change to the work.

4.4 Overview of Risk Management Process

The Code of Practice on Workplace Safety and Health (WSH) Risk Management (Workplace Safety and Health Council, 2015) describes the risk management (RM) processes that an organisation should implement (see Figure 4.5). As can be seen, risk assessment (RA) is a critical step of risk management (RM).

4.4.1 Preparation

The RM process starts with the formation of an RM team, or RA team if the workplace is relatively small and have less variety of activities

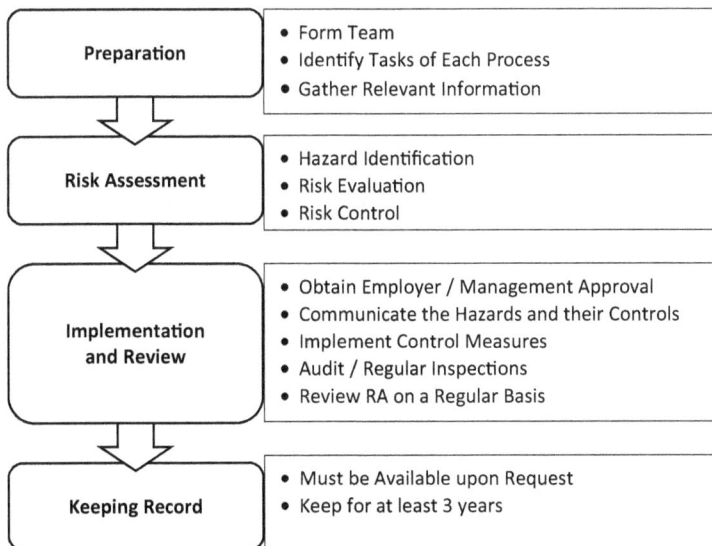

Figure 4.5   Risk management process (Workplace Safety and Health Council, 2015)

and hazards. A RM team consists of a RM leader (or champion) and multi-disciplinary team members who have knowledge of the different hazards and relevant controls. The RM leader or champion must have attended suitable RM training and be able to formulate a RM plan customised to the needs of the organisation. A RM plan is essentially a plan to implement the different controls so as to manage the WSH risk level of the workplace. Another important responsibility of the RM leader is to select the relevant team members with the right RM training, relevant expertise and suitable job responsibilities. The RM leader should have the ability to chair RA meetings effectively, i.e. he/she should have the ability to facilitate discussions, get members to contribute different opinions and help the group arrive at agreed decisions efficiently. The RM leader also needs to ensure proper documentation and record retention. Some possible team members include management staff, site engineers, technicians, safety personnel, supervisors, operators, contractors, and suppliers. The roles and responsibilities of the team members must be clearly established and only personnel trained in RA can be appointed as team leaders. For large organisations and organisations with a wide range of activities, there will be a need for RA teams to be created. Each RA team will be focused on specific activities or areas in the workplace and they will report to the RM team. The RA team leader will also be a member of the RM team. If the hazards and controls need specialist knowledge, or if the workplace does not have people versatile with RM/RA, it may be useful to engage consultants to assist with the process. However, the consultants should be there to assist and the RM/RA teams must still conduct the RM process and make the key decisions such as determination of risk levels and selection of controls.

The RM team will have to create an inventory of work activities and scope the RA. Scoping is essentially determining the boundaries for each risk assessment. This can be based on the activities, location, type of personnel, material, etc. The scope of RAs will also be guided by the organisation's policy, objectives, targets and programmes. For example, a developer may want to focus on hand and finger injuries arising from lifting work because based on their experience, those are some of the most frequent types of injury. In this case, an RA team can be specially formed and scoped to look into hand and finger injuries during crane lifting work. The RA will then form the foundation for risk controls such as a campaign to raise awareness on the prevention of hand and finger injuries during lifting.

During preparation, the RM/RA team will have to gather relevant information such as layout, workflow, manufacturer instructions, safety data sheet (for chemicals), etc. Different team members can assess different aspects of the collated information and then discuss them in meetings.

4.4.2 Risk assessment

After the preparation stage, risk assessment will have to be conducted. This stage includes hazard identification, risk evaluation and risk control. During hazard identification, the RM/RA team will have to brainstorm for possible hazards. There are many hazard identification methods. One of the most basic is to review relevant hazards or safety issues documented in WSH audit reports, inspection data, incident investigation reports, injury and illness records, occupational health surveillance reports, first aid logs, observation records, employee feedback or consultation records and safety data sheets. These documents will supplement systematic hazard identification or RA techniques like Job Safety Analysis (JSA) / Job Hazard Analysis (JHA), What-if Analysis, Failure Mode and Effect Analysis (FMEA), Hazard and Operability (HAZOP) Study, Fault Tree Analysis and Event Tree Analysis (see Table 4.2). The wide range of hazard identification and RA techniques can be confusing, but in the Singapore construction industry, most organisations use the approach recommended in the Risk Management Code of Practice (RMCP), which is essentially an activity-based RA (similar to JSA/JHA) or a trade-based RA. The activity-based RA and trade-based RA will be elaborated in this chapter later. In addition, if the workplace already exists, an inspection or walkabout can be conducted as part of the RA.

Table 4.2   Risk assessment techniques

Technique	Description	Remarks
Job Safety Analysis (JSA)/Job Hazard Analysis (JHA)	JSA and JHA are very similar techniques, both are conducted according to the following steps: • selecting the job to be analysed • breaking the job down into a sequence of steps • identifying potential hazards • determining preventive measures to overcome these hazards Some JSA requires the risk level to be determined, while JHA usually do not require risk levels to be determined. This is the key difference between the two techniques, but many variants exist.	JSA and JHA are relatively simple to use. They are usually structured based on a table similar to the table recommended in the RMCP. However, JHA does not require risk levels to be identified. They are suitable if the focus is on an activity, process or task that has specific steps. They are the most commonly used techniques in the construction industry.
What-if Analysis	A What-if Analysis consists of structured brainstorming to determine what can go wrong in a given scenario. The person performing the analysis then judges the likelihood that things will go wrong and considers the consequences. What-if Analysis can be applied at virtually any point in the evaluation process. Based on the answers to what-if questions, informed judgments can be made concerning the acceptability of those risks. A course of action can be outlined for risks deemed unacceptable. (adapted from https://www.acs.org/) The what-if questions should be generated prior to the brainstorming session and a group of experts must be involved in the analysis.	What-if Analysis is essentially a brainstorming session to identify hazards and assess their risk levels. People involved must be very experienced and the facilitator must be able to guide the group. As compared to the JSA/JHA approach, which is guided by a detailed template, the what-if analysis is comparatively unstructured.

Failure Mode and Effect Analysis (FMEA)	The FMEA is also a template-based approach, and it has a long history in reliability engineering. FMEA typically takes a piece of machinery or equipment and divides it into sub-systems that can be assessed effectively. The process involves: (1) Identification of the component and parent system, (2) Failure mode and cause of failure, (3) Effect of the failure on the sub-system or system and (4) Method of detection and diagnostic aids available. FMEA is quantitative in nature and it estimates the risk levels based on failure and error data collected by plants and manufacturers.	FMEA is commonly used in the manufacturing industry. Its quantitative nature makes it feasible only in situations where the data are readily available. Nevertheless, it is still possible to use it based on qualitative assessments.
Hazard and Operability (HAZOP)	Hazard and Operability (HAZOP) studies have been used for many years as a formal means for the review of chemical process designs. A HAZOP study is a systematic search for hazards which are defined as deviations within parameters that may have dangerous consequences. In the process industry, these deviations concern process parameters such as flow, temperature, pressure etc. (adapted from www.hse.gov.uk).	During a HAZOP study, the team will typically spend a few days to a few weeks evaluating the piping and instrumentation diagrams of an oil and gas plant to be constructed. The facilitator will use the set of guidewords (like more, less, high, low, fast, slow) to match with the process parameters so as to brainstorm what are the possible effects of these deviations. Common for industries like oil and gas, chemical and energy-related.

(Continued)

Table 4.2   (*Continued*)

Technique	Description	Remarks
Fault Tree Analysis and Event Tree Analysis	A fault tree is a diagram that displays the logical interrelationship between the basic causes of the hazard. Fault tree analysis (FTA) can be simple or complex depending on the system in question. Complex analysis involves the use of Boolean algebra to represent various failure states. (adapted from www.hse.gov. uk) While event tree analysis (ETA) is a forward, bottom up, logical modelling technique for both success and failure that explores responses through a single initiating event and lays a path for assessing probabilities of the outcomes and overall system analysis. This analysis technique is used to analyse the effects of functioning or failed systems given that an event has occurred. (adapted from https://en.wikipedia.org/wiki/Event_tree_analysis)	FTA and ETA are commonly used in the nuclear and process industry to analyse complex engineering systems. Human errors can also be included to determine the reliability of these complex systems. FTA and ETA are quantitative in nature but can be adapted to be used qualitatively.
Bowtie Analysis	The bowtie method is a risk assessment method that can be used to analyse and communicate how high-risk scenarios develop. A bowtie gives a visual summary of all plausible risk scenarios concerning a certain hazard that could occur. By identifying control measures, the bowtie displays what a company does to control those scenarios. Control measures are also known as barriers. They are elements we have in place to ensure the hazards we deal with are kept in a wanted state. Barriers can be systems, regulations, design aspects, and so on. (adapted from Bowtie Methodology Manual by CGE)	The bowtie is a qualitative method that relies on a set of symbols to summarise a series of hazards and controls into diagrammatic form. The bowtie requires specialised software to be implemented effectively as the number of symbols and diagrams will grow very quickly. However, each bowtie diagram is a good summary of the hazards and controls that a particular activity has.

During these walkabouts, checklists can be used to help RA teams identify hazards and observations, and interviews can also be conducted to obtain valuable information about how existing controls are performing. In general, the following categories of hazards should be considered:

- physical (e.g. fire, noise, ergonomics, heat, radiation);
- mechanical (e.g. moving parts, rotating parts);
- electrical (e.g. voltage, current, static charge, magnetic fields);
- chemical (e.g. flammables, toxics, corrosives, reactive materials);
- biological (e.g. blood-borne pathogens, virus); and
- psychosocial (e.g. stress, fatigue).

Human and cultural factors are important aspects that must be considered during RA. Some of the key factors include personal risk factors, e.g. decreased mental alertness, fatigue, loss of concentration. Another aspect is the tendency for workers to take shortcuts or violate safety rules, e.g. by-passing safety procedures, disabling machine safety features, not using or misusing personal protective equipment, and unauthorised use or misuse of equipment. Other at-risk behaviours include reckless acts and horseplay. Work organisation factors include excessive workload, prolonged working hours, lack of adequate training, and inadequate acclimatisation to hot work environment. Individual health risk factors, including medical health issues, smoking, and alcohol misuse, should also be considered.

The risk matrix in Figure 4.6 together with the severity levels and likelihood levels in Tables 4.3 and 4.4 respectively are described in the Risk Management Code of Practice (Workplace Safety and Health Council, 2015). They are used to evaluate hazards and assign a risk priority number (RPN). Once the RPN is determined, the risk level can then be determined, and the RM team can then refer to Table 4.5 for the recommended actions. Figure 4.6, Tables 4.3, 4.4 and 4.5, which are taken from the Risk Management Code of Practice (RMCP), are guidelines. Organisations can create their own risk matrix, severity and likelihood levels,

Likelihood Severity	Rare (1)	Remote (2)	Occasional (3)	Frequent (4)	Almost certain (5)
Catastrophic (5)	5	10	15	20	25
Major (4)	4	8	12	16	20
Moderate (3)	3	6	9	12	15
Minor (2)	2	4	6	8	10
Negligible (1)	1	2	3	4	5

RPN 1–3 = Low Risk; RPN 4–12 = Medium Risk; RPN 15–25 = High Risk

Figure 4.6   Risk matrix (Workplace Safety and Health Council, 2015)

Table 4.3   Severity level

Level	Severity	Description
5	Catastrophic	Death, fatal diseases or multiple major injuries.
4	Major	Serious injuries or life-threatening occupational diseases (including amputations, major fractures, multiple injuries, occupational cancers, acute poisoning, disabilities and deafness).
3	Moderate	Injury or ill-health requiring medical treatment (includes lacerations, burns, sprains, minor fractures, dermatitis and work-related upper limb disorders).
2	Minor	Injury or ill-health requiring first-aid only (includes minor cuts and bruises, irritation, ill-health with temporary discomfort).
1	Negligible	Negligible injury.

Table 4.4   Likelihood level

Level	Likelihood	Description
5	Rare	Not expected to occur but still possible.
4	Remote	Not likely to occur under normal circumstances.
3	Occasional	Possible or known to occur.
2	Frequent	Common occurrence.
1	Almost certain	Continual or repeating experience.

recommended actions, and calibrate them to reflect their risk appetite and tolerability. However, any deviation from the guidelines provided in the RMCP should be conservative, i.e. the appetite for risk would usually be decreased. One way to calibrate the matrix and levels is to identify common hazards that most employees of the organisation are familiar

Table 4.5   Recommended action for risk levels

Risk Level	Risk Acceptability	Recommended Actions
Low	Acceptable	• No additional risk control measures may be needed. • Frequent review and monitoring of hazards are required to ensure that the risk level assigned is accurate and does not increase over time.
Medium	Tolerable	• A careful evaluation of the hazards should be carried out to ensure that the risk level is reduced to as low as reasonably practicable (ALARP) within a defined time period. • Interim risk control measures, such as administrative controls or PPE, may be implemented while longer term measures are being established. • Management attention is required.
High	Not acceptable	• High Risk level must be reduced to at least Medium Risk before work starts. • There should not be any interim risk control measures. Risk control measures should not be overly dependent on PPE. • If practicable, the hazard should be eliminated before work starts. • Management review is required before work starts.

with. Next, and sample employees who are reflective of the population to assess those hazards individually. Subsequently, the RPNs generated by the sampled employees can then be used to assess if the definitions in Tables 4.3 and 4.4 are suitable. The higher the RPN, the more attention and control are required. The hazards assessed to be in the top right corner (where the RPN = 15 to 25) need to have additional risk controls before work can be conducted. Controls can be identified based on the hierarchy of control and Haddon's countermeasures. The diagonal area of the matrix (where RPN = 4 to 12) indicates that the risk level is tolerable (medium), but additional control measures should be implemented whenever reasonable and practicable. The lower left zone (where RPN

= 1 to 3) indicates that the hazard is low-risk and generally acceptable, but practicable controls that can push the risk lower should still be implemented. The different zones are aligned with the HSE Framework for Tolerability of Risk (Figure 4.3).

4.4.3 Implementation and review

To facilitate implementation, at the end of the RA, there should be a RM plan to ensure that all stakeholders are aware of the key hazards, their risk levels and the corresponding control. The RM plan should define the scope of the risk management, the company job functions, the RA methodology, the risk control measures and the schedule or programme for implementing the RM plan. Most importantly, the RM plan should contain the actions that are required to implement the different controls identified during the RA. Many a time, stopgap measures (usually lower on the hierarchy of control, i.e. administrative controls and PPE) need to be implemented first because more effective measures need time for implementation, e.g. installing an engineering control. These details need to be spelled out in the RM plan such that the risk level during operations is at most tolerable or medium. The plan should clearly stipulate:

- what controls have to be implemented
- how to implement the controls
- who are to implement the controls
- where to implement the controls
- when to implement the controls

During implementation, it is important for organisations to ensure buy-in from supervisors and workers through communication and consultation. In terms of communication, it is critical to ensure that the nature of the hazards, risk levels and controls were explained. The relevant safe work procedures should be communicated in a suitable format to the relevant stakeholders, e.g. in pictures or comics to overcome possible communication barriers. The controls will impact the work environment, equipment, work procedures and personnel. Thus, the impact of the changes should be clearly communicated before implementation.

For example, the RA team for the installation of a dry wall along the edge of a mezzanine floor in a building was in a rush to finish the RA. They identified the hazard of falling over the edge of the mezzanine floor (3 m above the ground floor) in a new workshop and noted that due to the lack of height clearance, lanyard and personal energy absorbers will not be suitable in protecting the workers. Thus, they recommended the use of safety or anti-fall nets to be installed along the perimeter of the mezzanine floor. However, the nets made it infeasible for the scissors lifts to be used to assist in the installation of the dry wall. Workers familiar with the work should have been able to identify the problem easily but were not consulted because the RA team conducted the RA in a rush. This resulted in delays in the project as a new RA had to be conducted and the installed safety net had to be removed.

To facilitate communications, the organisation should clearly establish the different channels of communications, including WSH committee meetings, feedback sessions, small group meetings (e.g. tool-box meeting, and shift handover meeting), one-on-one discussion, email or online messaging, telephone calls and notice board/bulletins. The stakeholders that must be involved in the development and communication of the risk management plan can include senior management, supervisors, subject matter experts, workers, consultants, contractors, and suppliers.

Effectiveness of the controls and risk of hazards should be monitored during implementation. Reviews should then be conducted at least once every three years, whenever there are significant changes to the work, or when there are incidents in the workplace. The RM plan should also clearly define the communication, consultation and review activities.

4.5 Problems and Challenges

Despite its importance, in actual implementation, RM and RA are saddled with numerous problems. As discussed earlier, risk perception is an important aspect of RM. Some of the cognitive biases affecting risk perception include (based on Dobelli, 2013):

o Base rate neglect — A disregard of fundamental frequencies of different types of accidents and illnesses. During RA, organisations fail to consider the overall accident frequencies in the industry and assign unrealistic likelihoods to their own workplaces.

o Over-confidence — When individuals are asked to rate their competency or likelihood to get into an accident, most will over-rate their competency in relation to others and assume that they will not get into an accident.

o Availability bias — Most people do not get into an accident or illness on a daily basis. Thus, based on their experience, the most available memory is those of non-accidents or days with no occupational illness. This can lead to biasness that accidents and illness will not happen on their site.

o Survivorship bias — Similar to the above biases, we tend to focus on success or 'non-failure'. "People systematically overestimate their chances of success. Guard against it by frequently visiting the graves of once-promising projects, investments and careers. It is a sad walk, but one that should clear your mind."

To deal with these biases, it is important to communicate WSH statistics and case studies to the managers, supervisors and workers. Awareness of these cognitive biases will also help to reduce their potential impact.

Another problem common in the industry is failure to implement the RA due to focus on productivity (rendering RA/RM a mere paper exercise) and poor communication. These issues are related to the safety culture of the organisation. Many organisations are over-focused on the documentations needed in RM. They think that having the set of documents means that they are protected from possible prosecution by the authorities. However, paperwork without implementation is still poor management and the paperwork takes up time without adding value. If accidents happen, the organisation will still be heavily punished.

Communication should always be focused on the receiver's preferences and suitability for the mode of communication. For example, the use of pictures and videos is much more effective for workers and supervisors than lengthy risk assessment documents. In the construction industry, where workers of different nationalities are present, the clarity of communication needs to be looked into very carefully. The person conducting the briefing must prepare beforehand, conduct the briefing clearly and check if the receivers of the information have understood the information. These basic communication processes are still very lacking in the construction industry.

Review Questions

1. Why is risk assessment the cornerstone of the WSH Act?
2. What are the factors that affect risk perception?
3. Describe the roles of the following people: Employer, Manager, Human Resource Manager, Risk Management and Risk Assessment Leaders, and Employees.
4. When can an organisation stop monitoring a specific hazard?
5. Describe six categories of hazards.
6. Describe three examples of human and cultural factors that should be considered in an RA.
7. Using the recommended 5 × 5 matrix, what are some of the high-risk, medium-risk and low-risk hazards in the construction industry? How do you justify your answer?
8. What do you think are some of the possible problems in implementing the Risk Management CP in a small worksite?

References

British Standards Institute (2009). "BS ISO 31000:2009 Risk management — Principles and guidelines." London.

Dobelli, R. (2013). *The art of thinking clearly*, Hodder and Stoughton, London.

Workplace Safety and Health Council (2008). *Technical Advisory for Working at Height*, WSHC, Singapore.

Workplace Safety and Health Council (2015). "Code of Practice on Workplace Safety and Health (WSH) Risk Management." WSHC, Singapore.

5 Design for Safety

The Event Causation Technique

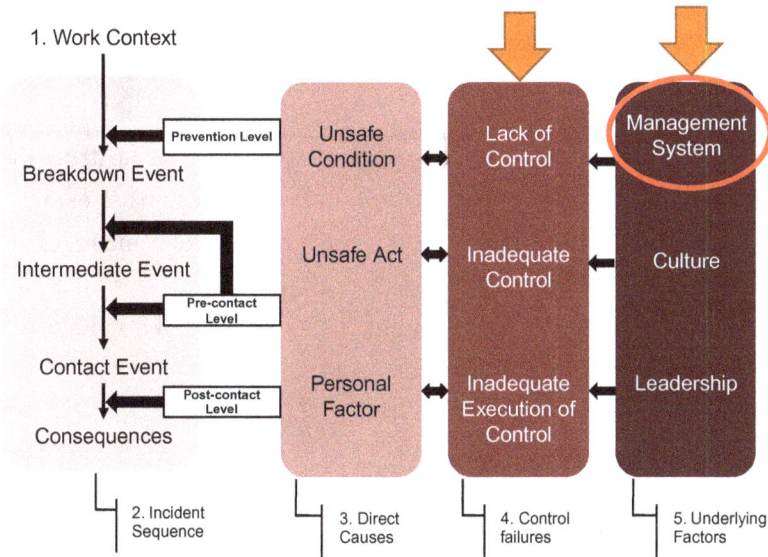

Design for Safety (DfS) is a design risk assessment. It is an important aspect of the management system (see Event Causation Technique (ECT) diagram above). The end products of the DfS process are the design-related controls to prevent incidents downstream.

5.1 Introduction

Design for Safety (DfS) promotes early consideration of safety and health hazards during the design phase of a construction project. With early intervention, hazards can be more effectively eliminated or controlled, leading to safer worksites and construction processes. DfS is practised in many countries, including Australia, the UK, and Singapore. In Singapore, the Ministry of Manpower (MOM) enacted the DfS Regulations in July 2015, which were implemented from August 2016 onwards.

The DfS Regulations are an implementation of one of the basic principles supporting the Workplace Safety and Health Act (WSH Act), "eliminate risk at source". Prior to the DfS Regulations, contractors have always been the main party involved in workplace safety and health (WSH). DfS imposes safety and health duties on the developers and designers, requiring them to conduct design risk assessments. Even though the DfS process is focused on WSH, it is also useful for the project team because it forces the project team to assess risks upstream so as to assure the effective implementation of the project downstream.

This chapter introduces the concept of DfS, relevant regulations and the Approved Code of Practice (ACoP). It also provides examples of how DfS can improve WSH throughout the lifecycle of a structure and discusses the challenges and success factors for DfS.

5.2 Defining Design for Safety

According to numerous studies, many fatalities in the construction industry can be attributed to design decisions or lack of planning (Behm, 2005; Workplace Safety and Health Council, 2015). Thus, the concept of DfS was introduced to minimise the risk of accidents and ill health through the consideration of hazards during upstream design phases of a construction project (Gambatese *et al.*, 2008). DfS is also known as prevention through design (López-Arquillos *et al.*, 2015), safe design (Safe Work Australia, 2012) and Construction (Design and Management) (Health and Safety Executive, 2015).

DfS is defined as: *"The practice of anticipating and "designing out" potential occupational safety and health hazards and risks associated with new processes, structures, equipment, or tools, and organizing work, such that it takes into consideration the construction, maintenance, decommissioning, and disposal/recycling of waste material, and recognizing the business and social benefits of doing so."* (Schulte *et al.*, 2008) (p. 115). Essentially, DfS is the risk assessment of a design to identify the hazards that can pose safety and health risks at different stages of development of a workplace, building or facility, so that the hazard can be eliminated or mitigated through options such as design changes, risk control measures or the sharing of information.

The DfS process requires the involvement of all stakeholders. In the construction industry, the stakeholders include the client or developer, the designers, and the contractors. Similarly, in other industries the owner or client will usually engage designers and contractors to design and manufacture or construct an asset, equipment, machinery or product. This asset, equipment, machinery or product can be a vessel, a metalworking plant, a school building, a photocopier or a desk. Some industries, for example the oil and gas industry, have a very rigorous process to ensure safe operation and construction. This process uses risk assessment methods like Process Hazard Analysis (PHA), Hazard and Operability (HAZOP), What-if Analysis and Layers of Protection Analysis (LOPA). The construction industry in Singapore also has a thorough design review process. The Singapore Building Control Act and Building Regulations provides stringent requirements to ensure that competent designers are engaged to oversee and check designs. However, the Building Control Act is focused on major hazards such as the collapse of a structure and building fires, and does not adequately cover WSH issues like slips, trips and falls, noise induced deafness and getting struck by vehicles. DfS is mainly concerned with WSH incidents that involve the actions of individual workers and is distinctly different from the focus of the Building Control Act. This chapter will discuss the concept of DfS using the construction industry, but the concepts are applicable to other industries as well.

The concept of DfS is not new to the construction industry. For example, the Europeans have been implementing the concept since the 1990s. In the UK, the Construction (Design and Management) (CDM) Regulations has been revised several times and had seen some successes. The South African Construction Regulations (2003) and the Australian

Occupational Health and Safety Acts have also implemented the concept of DfS. In Singapore, the DfS Regulations was enacted in 2015 and enforced in 2016. DfS is aligned with one of the key principles of the WSH Act, "reducing risks at the source by requiring all stakeholders to eliminate or minimise the risks they create." Despite the differences between the operationalisation of the concept of DfS, all countries implementing DfS seek to minimise risk of accidents and ill health through the consideration of hazards during the upstream design phases of a construction project.

As seen in Figure 5.1, as a project progresses from inception to completion, the scope for change decreases and the cost of change increases. Since DfS involves design changes to eliminate hazards and implementation of controls to mitigate WSH risks, as a project progresses, it becomes more difficult to implement these changes and controls to improve WSH. Therefore, it is important for DfS to be implemented early in the design phase. However, it is also not possible for DfS reviews to be conducted when there is insufficient information about the design. Thus, DfS reviews must be conducted near the end of each design phase, when it is still possible to implement changes. In

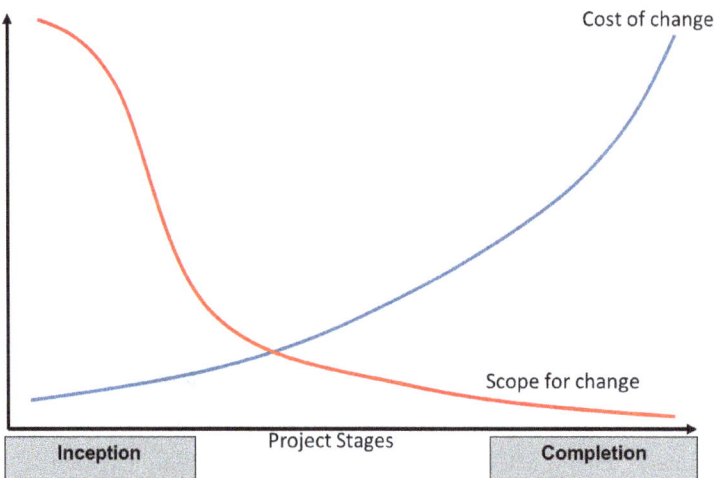

Figure 5.1   Cost of and scope for design changes to improve safety

addition, the DfS review process considers the safety and health of construction workers, occupants, people in the community, maintenance workers, and demolition workers. Thus, it is useful to involve construction contractors and facilities managers in the design review process to ensure comprehensive hazard identification. The DfS process will be discussed in more detail subsequently.

5.3 DfS Regulations

The following provides selected highlights of the Singapore DfS Regulations. The DfS Regulations are applicable to projects undertaken by a developer in the course of the developer's business, where the contract sum is S$10 million or more and involves development under section 3(1) of the Planning Act (Cap. 232). According to the Planning Act, development is defined as, "the carrying out of any building, engineering, mining, earthworks or other operations in, on, over or under land, or the making of any material change in the use of any building or land." The Planning Act section 3(2) describes specific exclusions that are not classified as development and hence not covered by the DfS regulations. For example, the maintenance, improvement or other alterations of a building which do not materially affect the external appearance or the floor area of the building, minor or preliminary works and such temporary use of land as may be declared by the competent authority, and carrying out by any statutory authority of any works for the purpose of laying, inspecting, repairing or renewing any sewers, mains, pipes, cables or other apparatus. On the other hand, the Planning Act section 3(3) highlighted several specific inclusions, e.g.

- "(b) the use as a dwelling-house of any building not originally constructed for human habitation involves a material change in the use of the building"
- "(c) the use for other purposes of a building or part of a building originally constructed as a dwelling-house involves a material change in the use of the building"

- "(d) the demolition or reconstruction of or addition to a building constitutes development"
- "(e) the use for the display of advertisements of any external part of a building which is not normally used for that purpose involves a material change in the use of the building"

However, it must be noted that the regulations only apply to projects that had a designer appointed on or after 1 August 2016. The DfS Regulations consist of four parts: (1) Preliminary, (2) Duties of Developer, (3) Duties of Designer and Contractor, and (4) Miscellaneous. The following section discusses the duties of the different stakeholders.

5.3.1 Developers

The developers are expected to ensure that "foreseeable risks" are eliminated. If it is not "reasonably practicable" to eliminate the risks, then the "design risk" must be "reduced to as low as reasonably practicable" by reducing the risk at its source and using collective protective measures instead of individual protective measures. In relation to a structure, "design risk" means anything present or absent in the design of the structure that increases the likelihood that an affected person (e.g. construction workers, and occupants who work in the structure after construction, including maintenance workers and demolition workers) will suffer bodily injury when constructing, working at, or demolishing the structure. The structure refers to any permanent or temporary structures and it includes the products and mechanical or electrical systems that are part of it. Some examples of structures can include a school building, a temporary earth retaining structure, a green façade on which a landscape worker works, and a warehouse that is being demolished. The concept of reasonably practicable has been discussed in Chapter 1.

An example of working at heights is used to illustrate what it means to reduce risk at source and what collective and individual protective measures are. A DfS review for a condominium building using volumetric construction methods identified that during construction, many construction workers need to work-at-heights to install the volumetric modules. Even though the risk is lower than traditional cast-in-situ methods, the team of installers and lifting personnel will still be exposed to the risk of falling from height. The current approach is to require the site workers to install their own fall arrest lifelines when the modules are being installed on site. This means that contractors will have to manage the risk downstream (not at source). The fall arrest system is an individual protective measure. Even though it is ideal to totally eliminate the risk of falling from height, it is not feasible with the current state of technology. Thus, following the principle of reducing risk at source and the use of collective protective measures, the design team decided that they will design a set of foldable guard rails that can be folded when the modules are being transported. The foldable guard rails can then be put into place when the modules are still at the ground level of the site. The worker can use platform ladders or scissors lifts to install the guard rails before they are being lifted. The risk is now greatly reduced because the risk control measure (foldable guard rails) are designed for upstream installation and put in place at the prefabrication yard and not at the construction site. The guard rails are a collective protective measure because, in contrast to a fall arrest system, it can protect many workers at the same time.

Developers are to ensure that designers and contractors are competent to perform their duties under the DfS Regulations. The traditional approach of ensuring competency is through training. Thus, developers will usually require designers and contractors to have personnel who have taken DfS-related and WSH-related courses. However, another way to demonstrate competency is through past experience in implementing DfS review processes.

In addition, developers have to ensure that the project is properly planned and managed. The DfS Regulations specifically highlighted the need for "sufficient time and resources" and "relevant information" to be provided to designers and contractors so that they can perform their statutory duties under the WSH Act. However, it is not clearly established what constitutes "sufficient time and resources". One possible approach to determine sufficient time and resources is to make reference to the time and resources allocated to past similar projects that were safely completed. More research is needed to provide benchmarks to evaluate the sufficiency of time and resources allocated.

Developers need to convene DfS review meetings to identify all foreseeable design risks and discuss how each of the risks can be eliminated or reduced. The DfS processes involved will be discussed in more detail subsequently. Developers will have to ensure that all relevant designers and contractors attend the DfS review meetings. A DfS register containing information and records of the DfS review meeting and residual design risk must be kept up-to-date and made available to designers, contractors and the regulator (i.e. WSH Inspectors from the Ministry of Manpower). The DfS register is a legal document that must accompany the structure even when the owner changes. The developer must inform any new owner of the nature and purpose of the DfS register.

The developer is allowed to delegate the duties related to DfS review meeting and DfS Register to a "DfS Professional" (DfSP), who is a person who is assessed by the developer to be competent to perform the duties. The WSH Council approved several industry associations, such as the Institution of Engineers Singapore (IES), the Singapore Institute of Architects (SIA) and the Singapore Contractors Association Ltd (SCAL), to conduct a two-day DfSP course to certify DfSPs. Participants are required to have significant experience in the construction industry. The course also requires participants to submit a report containing a DfS register for assessment.

The developer must provide all the relevant information necessary for the DfSP to perform the delegated duties. On the other hand, the DfSP must, as soon as reasonably practicable after each design-for-safety review meeting, provide the developer with all relevant information on each foreseeable design risk identified at the meeting and how each design risk can be eliminated or reduced. In addition, the DfSP must, as soon as reasonably practicable after any information or record is added to the DfS register, provide the developer with an updated copy of the

DfS register. It must be noted that the developer cannot delegate its general duties with regard to the implementation of DfS, ensuring competent designers and contractors are engaged, ensuring that the project is properly planned and managed with sufficient time and resources for designers and contractors, and ensuring designers and contractors have all relevant information.

From the duties allocated in the DfS Regulations, it is obvious that developers has an important role to play in DfS. This is a suitable approach because past research (Goh and Chua, 2016; Toh *et al.*, 2016) has shown that developers are an important source of motivation for designers to implement DfS.

5.3.2 Designers

In the DfS Regulations, a "designer" is defined as the person who prepares a design plan relating to a structure. "Design plans" include drawings, building information modelling, design details, specifications, materials and bills of quantities (including specifications of articles or substances) relating to a structure, and calculations prepared for the purpose of a design. This definition is very broad and can even include quantity surveyors and contracts managers who develop bills of quantities that influence the safety of the design.

While preparing the design plan, designers (in particular, engineers, architects, and contractors) must "as far as reasonably practicable eliminate all foreseeable design risks". If it is not possible to eliminate the design risk, then the designer must propose to the person who appointed the designer (e.g. developer, other designer or contractor) a modification to the design plan that reduces the design risk to as low as reasonably practicable. As discussed earlier, the modification must take into account the principle of reducing risk at source and adopt collective protective measures instead of individual protective measures. Designers must provide all relevant information on the design, construction or maintenance of the structure to the person who appointed the designer.

Many designers do not have sufficient experience and knowledge of the downstream processes, e.g. construction, maintenance and demolition. This knowledge gap can be minimised by involving construction contractors and facilities managers early in the process. Concurrently, it is useful to train designers on WSH hazards and controls, so that they are better able to assess WSH issues during the design phase.

5.3.3 Contractors

Contractors must inform the person who appointed them of any foreseeable design risk that they know of. They must also ensure the designers and subcontractors engaged by them are competent and provide them with relevant information.

Contractors are heavily regulated in terms of site WSH management and have the most direct control over site safety. In contrast, DfS is more focused on developers and designers. However, in the case of design and build projects, the contractor performs both the role of a designer and a contractor. As discussed earlier, it is advantageous for contractors to be involved in the design as early as possible so that they can provide inputs on the design before it is frozen. Contractors also engage designers to design temporary structures. In this context, they will also have to ensure that the designers are versatile with the DfS process.

5.4 DfS Process

This section discusses two approaches used to facilitate the DfS process, CHAIR and GUIDE, with attention on the GUIDE process detailed in the WSH guidelines — DfS (Workplace Safety and Health Council, 2016) (hereafter referred to as "DfS Guidelines").

5.4.1 CHAIR: Construction Hazard Assessment and Implication Review

The Construction Hazard Assessment and Implication Review (CHAIR) was created by WorkCover New South Wales (2001). CHAIR is essentially a customisation of the HAZOP study (see Chapter 4 for more detail) for the construction industry. CHAIR aims to identify and eliminate hazards or minimise risk levels of the hazards in a design as early as possible. The process involves all key stakeholders, who gather together to reflect on the design from a health and safety perspective. CHAIR is implemented at two key junctures in the project lifecycle: during the conceptual stage of a design (CHAIR ONE) and just prior to construction (CHAIR TWO and THREE). CHAIR ONE is meant to review the concept design. CHAIR TWO is meant to review construction and demolition issues. CHAIR THREE is focused on maintenance and repair issues.

CHAIR involves the following steps:

1. Assemble a CHAIR study team (include all stakeholders).
2. Define the objectives and the scope of the study.
3. Agree on a set of guidewords/prompts to assist brainstorming process.
4. Partition the design (CHAIR-1, CHAIR-3) or construction process (CHAIR-2) into logical blocks of appropriate size.
5. For each logical block, use various guidewords to assist with the identification of safety aspects/issues.
6. Discuss associated risks and determine if the safety risk can be eliminated.
7. If the safety risk cannot be eliminated, determine how it might be reduced.
8. Assess whether the proposed risk controls (e.g. expected safeguards, etc.) are appropriate (i.e. is the risk as low as reasonably practicable).
9. Document comments, actions and recommendations — determine appropriate solutions for design issues still to be resolved.

Since HAZOP is known for its rigour in assessing the safety and operability of a design, typically for process plants, CHAIR is also a rigorous process. However, HAZOP has not been known to be well-used in the construction industry due to the difficulties in creating guidewords and the time taken to conduct a HAZOP study. A brief enquiry with a major Australian construction contractor and an experienced Australian occupational safety and health expert in the construction industry did not indicate that CHAIR is well-used in Australia. In addition, Bluff

(2003) showed that the Memorandum of Understanding (MOU) signed by 17 contractors in New South Wales (NSW), Australia, which provided the support for the development of CHAIR, did not have a significant impact on design for safety in NSW. Bluff (2003) indicated that the challenges in implementing design for safety are: too much focus on paperwork, failure to address safety in design by clients, design profession and principal, poor programming practices and unrealistic timeframes. It can be hypothesised that the usefulness of CHAIR as a design for safety tool is highly dependent on the safety attitude of clients, designers and contractors, adequate planning and the provision of a reasonable timeframe. The introduction of new tools (e.g. CHAIR) and roles (e.g. Design for Safety Professional) into the design process requires understanding of the fundamental problems of the construction industry so as not to exacerbate the situation.

5.4.2 DfS guidelines by WSHC

The DfS guidelines (Workplace Safety and Health Council, 2016) describe a recommended process of design safety review and documentation. The overall design review process is based on the GUIDE acronym which represents:

1. **G**roup together a review team consisting of main stakeholders
2. **U**nderstand the design concept by looking at the drawings and calculations or have designers elaborate on the design
3. **I**dentify the hazards and risks that arise as a result of the design or construction method. The risks and hazards should be recorded and analysed to see if they can be eliminated by changing the design.
4. **D**esign around the hazards and risks identified to eliminate or mitigate the risk.
5. **E**nter all the information — including information on vital design change, that would affect safety and health or remaining hazards and risks to be mitigated — into DfS register

Hazard identification (step 3) and design change (step 4) should be iterated until it is not reasonably practicable to eliminate the hazards. At the end of the GUIDE process, the residual risk level of hazards should be recorded and signed off by the project team. The developer or the appointed DfS Professional (DfSP) will have to ensure that the DfS Review process is embedded into the project.

According to the WSH Guidelines for DfS (Workplace Safety and Health Council, 2016) (see Figure 5.2), the GUIDE process should be conducted at three different junctures, each with a different focus. The three design reviews are named GUIDE-1, GUIDE-2 and GUIDE-3. For a traditional design-then-build contract, a GUIDE-1 design review is conducted after the concept design is nearly completed, but still has room for amendments. It is meant to focus on the project's general location, traffic, type of buildings in the surroundings, and other general characteristics. If necessary, at each of the stages, more than one meeting should be conducted to review all the possible hazards arising from the design. GUIDE-2 is meant to focus on the detailed design, maintenance and repair. According to the DfS Guidelines, GUIDE-2 should review the hazards related to construction methods, access and egress and whether the design will create confined space or other hazards. Hazards during the maintenance and repair of the structure should also be reviewed. Even though GUIDE-2 should include information provided by the

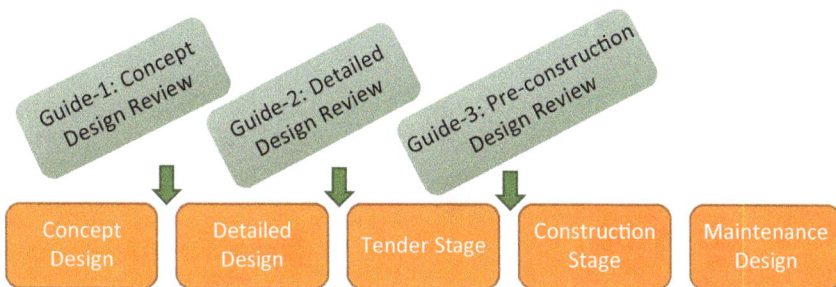

Figure 5.2   GUIDE process for design-then-build contracts (Workplace Safety and Health Council, 2016)

contractor, this will be challenging for a traditional design-then-build project as the contractor may not be appointed yet. However, in alignment with newer approaches such as early contractor involvement (ECI), contractors can be engaged as consultants to improve the design review process. Since hazards during the maintenance and repair phase will be identified and eliminated or mitigated, it is also useful to involve experienced facilities managers and maintenance contractors during GUIDE-2.

A design-and-build (D&B) contract (see Figure 5.3) has the advantage of having the contractor and lead designers integrated into one. This is aligned with the integrated design and delivery (IDD) approach which claims to produce better performing projects. In the case of a D&B contract, the contractor will be able to provide construction method information early and influence the design to eliminate or mitigate the risk of the construction stage.

GUIDE-3 is focused on the pre-construction review and the focus is on temporary works design and design by specialist contractors not covered during the concept and detailed design phases. Some of the key hazards and controls include: shoring, trenches and deep excavation, confined spaces and formwork and falsework. Residual risk of hazards and controls will have to be documented and highlighted to the contractors at the end of GUIDE-3. For example, if foldable guard rails are designed for the top of prefabricated modules to protect workers

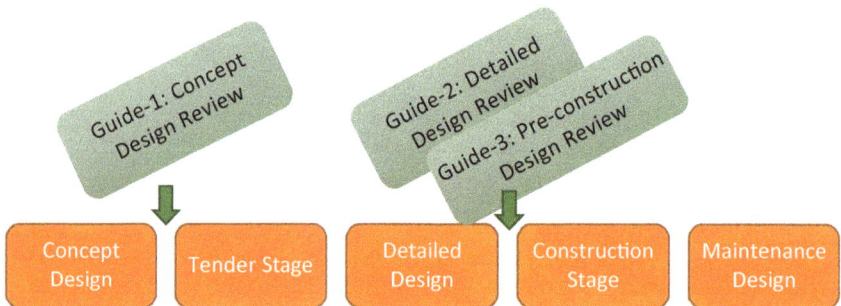

Figure 5.3 GUIDE process for design-and-build contracts (Workplace Safety and Health Council, 2016)

during installation, the folded guard rails need to be put into place by a site worker prior to the lifting of the modules. In this case, the residual hazard is falling from height when the worker is setting up the guard rails. The contractor will have to be notified of this hazard and will have to assess the risk during construction as well as put in place control measures to ensure that the risk is as low as reasonably practicable.

The DfS Guidelines highlight the following aspects that must be covered (but not limited to) during the GUIDE process:

• General design concept	• Layout
• Accessibility	• Maintenance
• Confined space	• Material handling or storage
• Emergency	• Means or methods
• Lighting	• Operation
• Excavation	• Physical hazards
• Fall prevention	• Sequence of construction
• Working platforms	• Standardisation of building elements
• Hoisting or weight	• Weather

The same concept of risk assessment used during the construction stage can be used during the GUIDE design reviews. However, the design review is usually organised based on locations or design elements instead of activities. The hazards identified will then be controlled through elimination or mitigation. If mitigation is selected, then the hierarchy of control will also be used to guide the design risk control. Since the designers can make changes during design, they are expected to adopt more substitution and engineering controls and rely less on administrative controls and personal protective equipment (PPE).

The DfS Guidelines also provide a Safety and Health Risk Assessment Form which can be used to guide the DfS reviews. A similar form is provided in the Appendix to this chapter. As with any risk assessment, the process must be structured and easy to administer. Since it is difficult to require all stakeholders to spend a lot of time in a DfS review meeting, it is important to predetermine a list of design considerations that should be discussed during the design review meeting. These design considerations can be identified through a preliminary evaluation conducted prior to the design review meeting through the use of checklists, the use of guidewords (see CHAIR method discussed earlier), and asking stakeholders to brainstorm using what-if questions on possible design-related hazards. The returns from all stakeholders will then be compiled by the developer or his representative (e.g. a DfS Professional). For example, the design considerations during a GUIDE-2 review can include air-con ledges, a skylight at the atrium area, a building maintenance unit, etc. These considerations can lead to hazards such as falling from height, getting struck by lightning and getting struck by falling objects. With the compiled list of design considerations and hazards, DfS review meetings can then be organised to focus on the issues identified.

It must be noted that the review should focus on hazards and controls not already covered in the Building Control Act and the Workplace Safety and Health Act. Since designers and contractors are already expected to comply with relevant regulations, the DfS review meetings should dedicate time to identify hazards peculiar to the structure being constructed. For example, if the construction site is next to a hospital, traffic to the hospital can be affected if the site entrance is placed along the road leading to the hospital. This can lead to emergency vehicles being delayed when trying to reach the hospital. In this case, even if having the entrance along the road to the hospital does not contravene any regulations, the entrance should be shifted. Compliance checks should still be conducted, but individual designers should be capable of doing that within existing building control processes.

Another area of focus during DfS reviews is the interface between design considerations and construction, maintenance and operational activities. DfS review meetings should consider the activities that would be occurring during the relevant activities related to relevant design considerations and consider the possible hazards that can arise. For example, during precast construction (i.e. design consideration is precast components), the delivery of precast components will mean the presence of trailers and an increased number of lifting operations. These activities will increase the risk of being struck by moving vehicles and being hit by objects falling from height. This will mean that during GUIDE-2 or GUIDE-3 the design of the site layout should consider ways to segregate site workers and delivery and lifting activities.

5.5 Examples of DfS

Readers should refer to the DfS Guidelines for samples and checklists for guidance on the GUIDE processes. Another useful source of examples of how designs can be amended to improve WSH can be found on the Design Best Practice website: http://www.dbp.org.uk/. The following are some useful examples:

1. Roof Access (http://www.dbp.org.uk/cs/DBP00061.pdf)
2. Piling safety (http://www.dbp.org.uk/cs/DBP00177.pdf)
3. Modular stone panels (http://www.dbp.org.uk/cs/DBP00009.pdf)

In addition, a series of case studies can be found in the compilation of worked examples in Hong Kong (Environment, Transport and Works Bureau (ETWB) *et al.*, 2003).

5.6 Challenges and Success Factors

According to Goh and Chua (2016), the top three perceived problems in practising DfS are client's cost concerns, contractor coming into the project too late, and inconsistency in DfS review or checks. To address these problems, the client needs to provide the necessary motivation to ensure that DfS is effectively implemented. As discussed earlier, early contractor involvement and integrated design and delivery (IDD) will be useful in improving DfS. Concurrent use of virtual design and construction (VDC) tools will help designers better foresee possible hazards.

Each DfS review will need to be effectively implemented. The following success factors for a DfS review are:

- Committed stakeholders who come to the DfS review meeting prepared and ready to contribute;
- An effective facilitator and coordinator for the DfS review process to ensure that all stakeholders are able to contribute to the review process;
- The design, WSH and operational competency of stakeholders (not just the facilitator) so that hazards, elimination and mitigation controls can be effectively identified;
- Information is available and reviewed prior to the meeting; and
- Detailed documentation and tracking of follow-up actions by stakeholders.

At the end, the personnel overseeing site operations (e.g. the Qualified Person in the Building Control Act and facilities managers) will have to make sure that the DfS measures are effectively implemented on site and during operations of the structure.

Review Questions

1. What are the roles of the different stakeholders in DfS?
2. Explain the GUIDE process and the differences between GUIDE 1–3.
3. Give examples of how design changes can improve: (i) WSH during construction, (ii) WSH during maintenance, (iii) WSH during use of facilities and (iv) WSH during demolition.
4. If you are the DfS Professional appointed by the developer, how would you ensure that a DfS review meeting is effective?

Appendix

Project Title: Company: Review date: Next review date:
Conducted by:
Process/ Location:

S/No.	Design consideration	Hazard identified	Risk assessment			Design out?	Proposed control	Residual risk level			Further review required	Action by (date)
			Severity	Likelihood	Risk level			Severity	Likelihood	Risk level		

References

Behm, M. (2005). "Linking construction fatalities to the design for construction safety concept." *Safety Science*, 43(8), 589–611.

Bluff, L. (2003). "Regulating Safe Design and Planning of Construction Works — A review of strategies for regulating OHS in design and planning of buildings, structures and other construction projects." National Research Centre for OHS Regulation, Sydney.

Environment, Transport and Works Bureau (ETWB), Hong Kong Housing Authority (HKHA), and Occupational Safety and Health Council (OSHC) (2003). "Construction Design and Management — Worked Examples." <http://www.devb.gov.hk/filemanager/en/content_29/cdm-worked%20 example.pdf>. (6 July 2017).

Gambatese, J. A., Behm, M., and Rajendran, S. (2008). "Design's role in construction accident causality and prevention: Perspectives from an expert panel." *Safety Science*, 46(4), 675–691.

Goh, Y. M., and Chua, S. (2016). "Knowledge, attitude and practices for design for safety: A study on civil & structural engineers." *Accident Analysis & Prevention*, 93, 260–266.

Health and Safety Executive (2015). "The Construction (Design and Management) Regulations 2015." <http://www.hse.gov.uk/construction/cdm/2015/ index.htm>. (22 July 2015).

López-Arquillos, A., Rubio-Romero, J. C., and Martinez-Aires, M. D. (2015). "Prevention through Design (PtD). The importance of the concept in Engineering and Architecture university courses." *Safety Science*, 73, 8–14.

Safe Work Australia (2012). *Safe design of structures*, Safe Work Australia, Canberra.

Schulte, P. A., Rinehart, R., Okun, A., Geraci, C. L., and Heidel, D. S. (2008). "National prevention through design (PtD) initiative." *Journal of Safety Research*, 39(2), 115–121.

Toh, Y. Z., Goh, Y. M., and Guo, B. H. W. (2016). "Knowledge, Attitude and Practice of Design for Safety: A Study on Multiple Stakeholders in the Construction Industry." *J. Constr. Eng. and Manage. — Am. Soc. of Civ. Eng.*, 143(5), 04016131.

WorkCover New South Wales (2001). "CHAIR — Safety in design tool." WorkCover New South Wales, Sydney.

Workplace Safety and Health Council (2015). "Design for Safety." <https://www.wshc.gov.sg/>. (12 Oct 2015).

Workplace Safety and Health Council (2016). "Workplace Safety and Health Guidelines — Design for Safety." <https://www.wshc.sg/files/wshc/upload/cms/file/WSH_Guidelines_Design_for_Safety(1).pdf>. (6 July 2017).

6

Overview of Workplace Safety and Health Management Systems

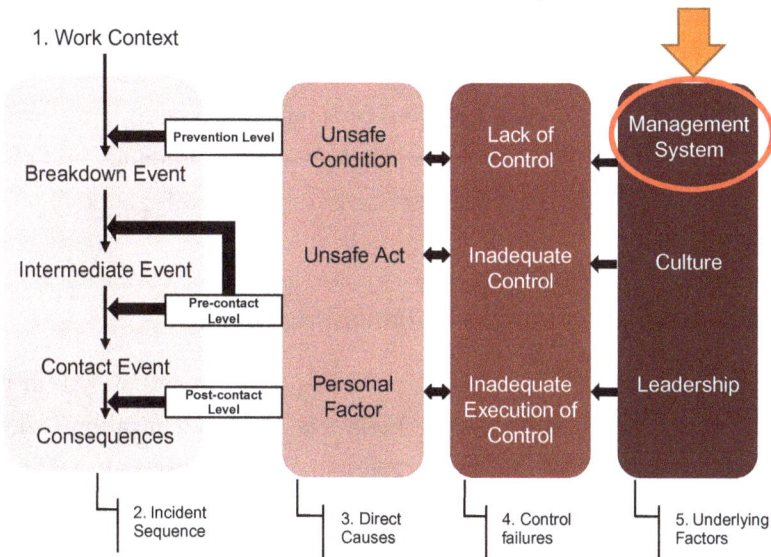

The Event Causation Technique

As identified in the Event Causation Technique (ECT) diagram above, the workplace safety and health (WSH) management system is an important underlying factor that can ensure that suitable risk controls are implemented effectively in response to incidents.

6.1 Introduction

This chapter provides an introduction to workplace safety and health (WSH) management systems. ISO 45001:2018 (ISO, 2018) will be used as an example of a WSH management system framework. It must be noted that there are other WSH management system frameworks, for example, CP 79: 1999 (Singapore Standards Council, 1999), ANSI/ASSE Z10-2012 9R2017) (American National Standard (ANSI)/American Society of Safety Engineers (ASSE), 2017) and the ILO Guidelines on occupational safety and health management systems (International Labor Organization (ILO), 2001). There are also numerous WSH management system frameworks developed by different companies such as ExxonMobil's Operations Integrity Management System (OIMS) (ExxonMobil, n.d.), and DNVGL's International Sustainability Rating System™ (ISRS) (DNVGL, n.d.). Many of the WSH management system frameworks developed by companies cover more than WSH. Other areas such as the environment, quality and security are frequently integrated together with WSH.

6.2 Management System as An Underlying Factor

According to definition 3.10 of ISO 45001, a management system is a "set of interrelated or interacting elements of an organization (3.1) to establish policies (3.14) and objectives (3.16) and processes (3.25) to achieve those objectives". The numbers in parentheses refer to corresponding definitions in other clauses in the standard. The term *elements* include "the organization's structure, roles and responsibilities, planning, operation, performance evaluation and improvement." Based on this definition of a management system, an occupational health and safety (OH&S) management system is defined as a "management system or part of a management system used to achieve OH&S policy", while OH&S policy refers to "intentions and direction of an organization (3.1), as formally expressed by its top management (3.12)".

As can be seen, a management system is essentially a set of fundamental management processes and structures. The overall purpose of a management system is to help an organisation set and achieve its policies and goals, and continually improve itself. The basic Plan–Do–Check–Act (PDCA) cycle is fundamental to a management system and one of the key purposes of the different elements is to help managers execute and maintain the PDCA cycle.

It is important to understand the difference between a management system and a specific control measure. A control can be a training course, specific personal protective equipment (PPE), or a safe work procedure. This will typically manage the direct causes of unsafe acts or unsafe conditions and these are usually event level remedies (see discussion in Chapter 4) that stand in contrast to a management system intervention, which focuses on the more fundamental elements of a management system, as highlighted earlier. Control measures are usually identified through risk assessments and they are meant to reduce the risks posed by a hazard. In contrast, a management system influences the patterns and trends of unsafe acts and conditions in the workplace.

For example, a worker works unsafely at height on wooden planks that are too flexible and unstable. An event level answer to the problem is that the planks are unsafe, change them and the worker is safe. However, a more systemic approach would be to identify that the planks are unsafe, determine if the planks are widely used in the organisation and whether workers are able to identify the danger that the planks pose to workers working on them. If the planks are used on all the sites in the organisation and workers are not able to perceive the planks as hazards, there is a dangerous pattern indicating that management system elements, for example, the risk assessment process (ISO 45001:2018 clause 6.1.2), procurement process (clause 8.1.4), and evaluation of training effectiveness (clause 7.2) may be inadequate. Using the procurement process as an example, it may be discovered that the procurement process did not include WSH criteria and feedback from site personnel were not included prior to the procurement of material. The procurement department may have been overly focused on price, and may not be able to account for safety criteria. Experienced site personnel should be required to provide input prior to purchase. The procurement process can then be improved by including a requirement for experienced site personnel to feedback on purchase specifications. A feedback form can also be developed, and relevant site personnel should be identified as relevant subject matter experts for different types of purchase.

The above example shows that by improving the management system elements, in this case the procurement procedure, the organisation as a whole becomes safer — not only will planks be safer, all newly purchased equipment should be safer due to input from experienced site personnel. Thus, with reference to the Event Causation Technique (ECT), a management system is an important underlying factor that interacts with culture and leadership to influence the effectiveness of all WSH controls and direct causes that can lead to incidents.

6.3 Overview of ISO 45001:2018

The following will provide an overview of ISO 45001:2018 as an illustration of the key elements of a WSH management system. Readers are encouraged to read the actual document for more details.

Figure 6.1 shows the management system framework of ISO 45001:2018. This framework is aligned with other ISO management system standards such as ISO 9001 (quality) and ISO 14001 (environmental management). Each of the elements in the framework will be discussed in the following sections. It is noted that the "initial review" was inserted by the author.

6.3.1 Initial review

With or without a formal management system, organisations have their own way of managing themselves. These processes may or may not be documented, but they are happening. When an organisation is trying to set up a formal and documented management system, they will have to first conduct an initial review, which compares existing management processes with the processes stipulated in the management system standard that they are trying to be certified to or model after, e.g. ISO 45001 or CP79:1999. As part of the initial review, existing risk management documents will have to be reviewed and additional risk assessments may

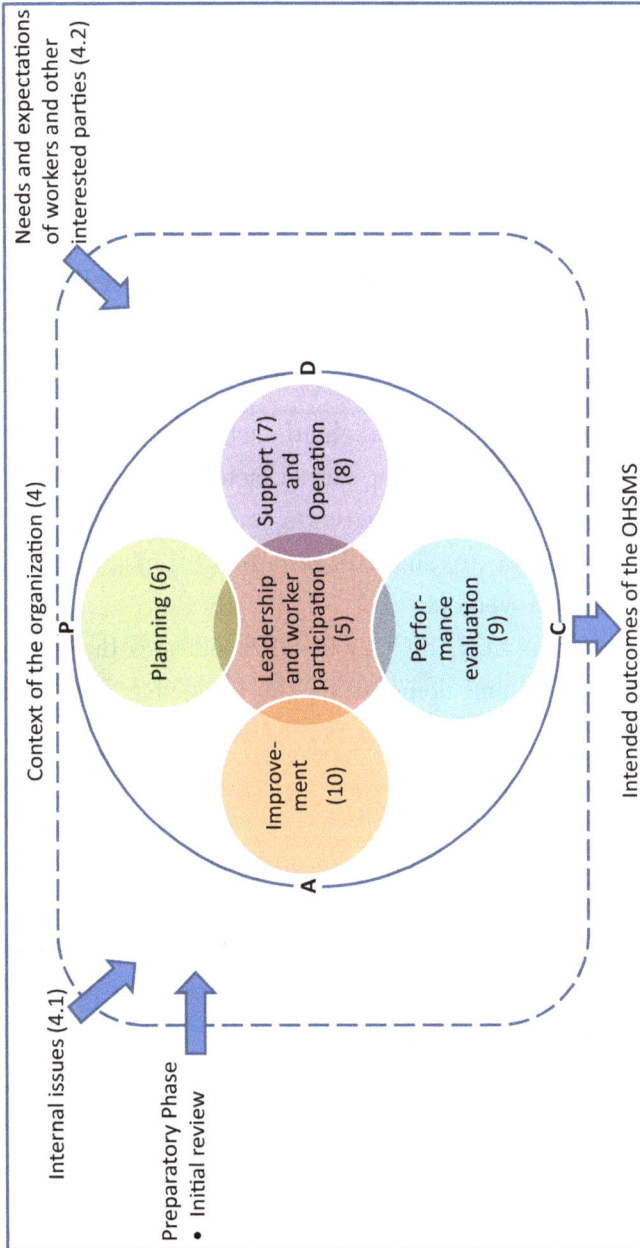

Figure 6.1   ISO 45001:2018 Management system framework; adapted from ISO (2018)

have to be conducted to understand the WSH risks that the organisation faces. A thorough check against legal requirements will also have to be conducted to understand the demands of the relevant legislations. Existing WSH trainings, programmes and initiatives should be documented. The resources used to conduct existing management system processes, programmes, etc. will also have to be clearly documented. Once the current management approach is documented, it should be compared against the WSH management system standard to identify gaps, so that specific actions can be identified and implemented to close the gaps.

6.3.2 Success factors

The ISO 45001:2018 clause 0.3 highlighted a set of success factors for an effective WSH management system. The 11 factors are classified into six categories: top management (sub-clauses (a) and (b)), policies ((f), (i), (j) and (k)), communication ((c) and (d)), resources (e), risk management (g) and continual improvement (h).

Aligned with the ECT, ISO 45001:2018 indicates that top management must provide and demonstrate the leadership, commitment, responsibilities and accountability for WSH management. The top management is responsible for developing, leading and promoting a positive safety culture that supports the prevention of incidents and mitigation of WSH risks. The top management must ensure that the WSH policies and objectives are compatible with the overall objectives and direction of the organisation and take into account the organisation's hazards, risks and opportunities. At the same time, for the WSH management system to be effective, it must be integrated into the organisation's business processes, and the management system must comply with relevant legal requirements and other requirements (e.g. customer requirements and industry standards).

Communication and consultation with the relevant interested parties will also help to ensure the success of the management system. Among the key interested parties are the workers and employees of the organisation.

The participation of workers in the development, implementation and maintenance of the management system will ensure that the management system is effective.

Top management needs to ensure that adequate resources are allocated to implement and maintain the management system to prevent the system from becoming a paper exercise. In addition, two other success factors highlighted in ISO 45001:2018, which are more process driven, are effective processes for risk management, and the continual performance evaluation and monitoring of the management system. Risk management is at the core of a WSH management system because through it hazards are identified, evaluated and controlled. Continual improvement processes ensure that the management system's inadequacies and opportunities for improvement are identified and actions are taken to perfect the WSH management system.

This set of 11 success factors is a useful reminder to managers of what to focus on when managing a WSH management system.

6.3.3 Context

A WSH management system does not exist in a vacuum. The context that it exists in contains different issues that will affect the way that the WSH management system should be designed, operated and maintained. These issues can be classified into internal and external issues. BS 45002:2018 (BSI Standards Limited, 2018) provides further guidelines on what are some of the possible internal and external issues. External issues include relationships with external providers such as contractors or suppliers, new technologies, cultural, social and political factors, and legislations. Internal issues include the size, nature and activities of the organisation, the way the organisation is managed and its business objectives, resources, knowledge and competence, and planned or foreseeable changes and how these are managed.

BS 45002:2018 (BSI Standards Limited, 2018) recommended approaches such as "what if" questions for organisations that run less complex operations and are smaller in size. For other organisations, structured methods such as SWOT (Strengths, Weaknesses, Opportunities and Threats) or PESTLE (Political, Economic, Social, Technological, Legal, Environmental) analysis can be used.

One key internal context that must be clearly established are the needs and expectations of workers and other interested parties. Thus, clause 4.2 of ISO 45001:2018 requires the organisation to identify workers (including managerial and supervisory staff) and other interested parties that are relevant to the WSH management system, their needs and expectations, and the relationships between these needs and expectations and relevant legal requirements.

Clause 4.3 specifies that the scope of the WSH management must be determined. The scope, i.e. boundaries and applicability, can be determined based on the types of activities, locations, products and services that the organisation has. For example, an organisation may choose to exclude certain locations from the WSH management system because the location is regulated by some prescriptive legal requirements that make the processes in the WSH management system unsuitable.

6.3.4 Leadership and worker participation

6.3.4.1 *Leadership and commitment (Clause 5.1)*

Unlike its predecessor, OHSAS 18001:2009, ISO 45001:2018 places additional attention on leadership and worker participation. Clause 5.1 stipulates 13 sub-clauses on the top management's responsibilities. These includes taking overall responsibility and accountability for WSH, ensuring the establishment of the WSH management system, integrating the WSH management system into the organisation's business processes, ensuring adequate resources, communicating the importance of effective WSH management and conformance, etc. The responsibilities are essentially elaborations of the 11 success factors highlighted earlier as well as the responsibilities over processes highlighted in other parts of ISO 45001:2018.

6.3.4.2 *WSH policy (Clause 5.2)*

The policy is one of the most fundamental parts of a WSH management system. It is essentially a statement that top management issues to demonstrate their commitment. However, despite its theoretical importance, the policy statement does not have a direct impact on an organisation's WSH, and it is becoming too easy for organisations to simply employ consultants to help them draft suitable policy statements. The most important part of Clause 5.2 is perhaps the phrase "[to] establish, implement and maintain" — especially the word "implement". It is important for the policy, which is the key guidance for the rest of the management system, to be implemented. To be implemented means that the management system does not just exist on paper but is actually being used by people on the ground. "Maintained" means it is being utilised, checked and improved in a sustainable manner. This requires active effort on the part of the organisation. Many systems start well, but deteriorate due to lack of maintenance. Many of the elements of ISO 45001:2018 (such as performance evaluation and improvement) are designed to ensure the active maintenance of the system.

6.3.4.3 *Organisational roles, responsibilities and authorities (Clause 5.3)*

The top management is required to assign responsibilities and authorities for relevant roles in the WSH management system. This can include roles such as a management representative overseeing the WSH management system, risk assessment teams, WSH officers, WSH committee members and WSH internal auditors. The roles, responsibilities and authorities must be documented and communicated to all interested parties. Workers (including managers and supervisors) shall assume responsibility for those aspects of the WSH management system over which they have control. However, ISO 45001:2018 clearly states that top management is still accountable for the functioning of the WSH management system.

One of the most basic ways to assign roles, responsibilities and authorities is to have an organisation chart and accompanying descriptions of the relevant WSH roles, responsibilities and authorities for each position in the organisation chart. In addition, for each process or project, the roles and responsibilities can be described based on "RACI", which stands for Responsible, Accountable, Consulted and Informed. "Responsible" refers to managers or persons-in-charge of the actual implementation of the work or process. "Accountable" refers to the owner or champion for the work or process who signs-off or approves the work or process. The person accountable will take overall responsibility for the work or process. "Consulted" refers to interested parties that need to give inputs to the work or process to enable effective implementation. And "Informed" refers to people that need to be kept in the picture, but do not need to be formally consulted and that are not directly involved in the work or process.

6.3.4.4 *Consultation and participation of workers (Clause 5.4)*

ISO 45001:2018 requires (a) process(es) for the consultation and participation of workers (and their representatives) at all levels and functions. Participation is defined as involvement in decision-making, while consultation is defined as seeking views before making a decision. The consultation and participation mechanisms must be supported by the necessary resources, training and time. In addition, management must determine and remove or minimise obstacles or barriers to participation. Some of these barriers and obstacles include failure to respond to worker inputs or suggestions, language or literacy barriers, reprisals or threats of reprisals, and policies or practices that discourage or penalise worker participation.

ISO 45001:2018 emphasises the consultation of non-managerial workers on issues such as WSH needs and expectations, WSH policy, assignment of roles and responsibilities, WSH objectives and plans, controls for outsourcing, procurement and contractors, WSH monitoring, measurement and evaluation, audit and continual improvement. In terms of the participation of non-managerial workers, the following are emphasised: mechanisms for consultation and participation, risk assessment and controls, competency issues, communication processes, incident investigation and corrective actions.

6.3.5 Planning (Clause 6)

The organisation must create plans to achieve the policy and the desired outcomes of the WSH management system, i.e. prevention of incidents and reduction of WSH risks. The planning element consists of actions to address WSH risks, opportunities (clause 6.1) and objectives, and planning to undertake them (clause 6.2). Clause 6.1 is focused on the risk assessment process, which is similar to the process covered in Chapter 4 on risk management. The management system will have to specify the procedure for the risk management process. In the context of Singapore, the risk assessment procedure will take reference from the Risk Management Code of Practice and other national guidelines. It should be noted that there are some differences in the terminology used in ISO 45001:2018 and the Risk Management Code of Practice (RMCP). For example, in the RMCP, risk assessment refers to hazard identification, risk evaluation and risk control, but in ISO 45001:2018, risk assessment is not clearly defined, and hazard identification and assessment of risks are assessed separately. Putting the differences in terminology aside, both emphasise the importance of conducting a systematic risk assessment (hazard identification, risk evaluation and risk control). It must be noted that ISO 45001:2018 (clause 6.1.2.3) highlights the need to have process(es) in place to assess WSH opportunities (in contrast to hazards and risks) to enhance WSH performance. The standard also requires considerations such as design of work areas, processes, installations, machinery/equipment, operating procedures and work organisation, which overlap with the concept of Design for Safety, which is covered in Chapter 5.

Clause 6.1.3 highlights the importance of having a procedure for identifying and accessing the legal and other Safety, Health and Environmental [management] (SHE) requirements that are applicable to the organisation. This may seem straightforward, but due to the wide range of applicable legislations and standards, it is actually a difficult task to keep updated on the changes and evaluate the possible impact of these legislations and standards. Thus, many organisations subscribe to information services to keep themselves updated.

Clause 6.1.4 requires organisations to plan for actions to address risks, opportunities, and legal and other requirements, as well as to prepare for and respond to emergency situations (clause 8.2). The plan must also integrate and implement the actions into relevant processes and evaluate the effectiveness of the actions.

Clause 6.2 is concerned with objectives. Policies identify issues of great importance to the organisation; based on the policy statement and WSH plans, objectives can then be created. For instance, if the organisation is concerned with the protection of the ozone layer, then they can set an objective to eliminate the use of chlorinated solvents, which are used in degreasing and cleaning activities but are also carcinogenic, in all manufacturing lines by 2016. The objective can also be supported by more specific targets, e.g. to replace 50% of chlorinated solvents in all manufacturing lines by June 2016 and to replace 75% of chlorinated solvents by Sep 2016. Once a set of objectives and targets has been determined, the organisation can have a plan, which identifies:

(a) what will be done;
(b) what resources will be required;
(c) who will be responsible;
(d) when it will be completed;
(e) how the results will be evaluated, including indicators for monitoring;
(f) how the actions to achieve OH&S objectives will be integrated into the organisation's business processes.

A series of activities, campaigns and events that help the organisation achieve its policy, objectives and targets will be contained in the plan. The plan must have a clearly identified timeline and allocation of responsibilities and accountabilities.

6.3.6 Support (Clause 7)

To support the functioning of the different elements of the WSH management system, ISO 45001:2018 identified five areas of support: resources, competence, awareness, communication and documented information. Resources are relatively self-explanatory and have been highlighted in some of the earlier clauses. Competence is a critical area in WSH and workers (including managers and supervisors) need to be trained so that they can perform their work safely. Competence goes beyond mandatory training required by the government. Mandatory training often only provides the basics and on-going customised training is more critical. Thus, a training needs analysis must be conducted based on the risk assessments, applicable legislations, etc. Besides training, experience is also critical in ensuring competence.

Besides competence, workers need to be aware of a range of WSH management system information such as the policy and objectives, benefits of improved WSH performance, consequences of poor WSH, incidents and investigation outcomes, risk assessment outcomes and actions, and their right to stop work when the WSH risk is not acceptable.

Communication refers to the communication processes needed for internal and external WSH communications. The processes must include determination of:

(a) What needs to be communicated;
(b) When to communicate;
(c) With whom to communicate (internal, contractors, and interested parties); and
(d) How to communicate.

The communication processes must take into account possible communication barriers like literacy levels, language and cultural diversity, and disabilities.

Documentation is a fundamental process in a formal management system, but the level of documentation must be proportional to the level of complexity and size of the operations. Legal requirements are also a major factor to consider when considering documentation. At the same time, the control of documented information needs to be robust to ensure the accessibility and reliability of the information. A common problem in organisations is the failure to ensure proper versioning of documents, leading to out-dated documents being used at the workplace. Processes for document control are critical to prevent miscommunication and application of out-dated procedures.

6.3.7 Operation (Clause 8)

Clause 8 covers the processes that are needed to prevent WSH incidents and this relates to the plans and actions identified during planning. Operational planning and control covers elements such as determining necessary process criteria that define safe operations, safety work procedures, and work instructions. The controls are identified based on the risk assessment, and those that are similar across different risk assessments can be grouped together as general control measures, e.g. basic WSH rules, hazardous chemical procedures, programmes for the maintenance and repair of facilities, housekeeping procedures, traffic management plans, occupational health programmes, permit-to-work systems and personal protective equipment programmes. High risk activities that present specific hazards will require a dedicated set of controls and procedures for the activities based on the hierarchy of control discussed in the earlier chapter on risk management.

An important process during operation is management of change (MOC) (clause 8.1.3). MOC is implemented whenever there are changes in the workplace that influence WSH. These changes can be temporary (e.g. a truck breaking down at a construction site during a casting operation) or permanent (e.g. a new pressure vessel on the factory floor). It can also be due to external issues like new legislation or technology. The organisation needs to define processes that evaluate these changes, assess the WSH risk and opportunities, and implement controls to manage the risk and opportunities.

Procurement (clause 8.1.4) is focused on the establishment of WSH requirements for products and services to be purchased. The purchasing department will have to identify the standards that purchases have to comply with, write up relevant WSH specifications such as required certifications, expectations of the safety performance of the service provider and required audits on the service provider or supplier. The procurement department will have to inform the supplier or service provider on what is expected of them in terms of their WSH management systems. The purchasing organisation may expect the supplier to be involved in the WSH activities of the organisation, e.g. participate in safety campaigns, have a similar level of training and similar risk assessment. Pre-approvals of requirements, specifications and procedures for the purchase of chemicals, machinery, and equipment will require the involvement of relevant experts and a clear process for sufficient checks prior to purchase. The WSH performance of suppliers and the goods that they have supplied should also be recorded and evaluated. This information can then be used for future purchase evaluation. Goods, equipment and services should be inspected and verified in terms of the WSH specifications and documented for future purchases.

Another important area is control over contractors. Many organisations now use many contractors in their work. There are advantages in using contractors with the expertise that the organisation lacks, but organisations need to manage the contractors carefully. Contractors can be unfamiliar with the work environment, WSH requirements and hazards of the client organisation, and they may have a weak WSH management system. Thus, it is important to establish clear criteria to select contractors who can demonstrate their capability to manage WSH. To do this, many clients rely on WSH certifications to standards like ISO 45001:2018 and other safety schemes or standards. In Singapore, experience shows that relying on certifications and other schemes like bizSAFE[1] may not be the best gauge of good WSH performance. A mix of information like interviews, audit reports, client's own experience with

---

[1] A scheme under the Workplace Safety and Health Council, which is described as "a five-step programme that assists companies to build up their WSH capabilities so that they can achieve quantum improvements in safety and health standards at the workplace."

the contractor, and references from other clients are critical information in the contractor selection process. Accident and incident records are useful sources of information, but many companies are simply lucky and serious accidents may be too rare to provide a good indicator. Nevertheless, accident and incident records must be reviewed when selecting contractors. Contractors, especially term contractors or those dealing with larger projects, will have to be evaluated and reviewed for their WSH performance and diligence in implementing their WSH management systems. Some basic checks would include on-going inspections, review of method statements and risk assessments, and review of safety and health committee meeting minutes. An important aspect is to determine how senior management and line management view WSH. If they see it as the job of the WSH department, then it is an indication that WSH is not taken seriously.

Suppliers and contractors are critical stakeholders in an organisation's management of WSH performance. Contractors frequently deal with highly hazardous processes. Due to the short-term nature of work, varying levels of risk tolerability and other factors, contractors can cause WSH incidents. Clients, through their contracts or WSH department, will have to monitor WSH management activities of the contractors from selection to evaluation and monitoring. It is critical to store organisational knowledge of different contractors and suppliers so that poor performing contractors and suppliers will not be re-engaged. However, it is important to take a partnering approach, where the client helps the contractors and suppliers improve and use incentives and recognitions to motivate the contractors and suppliers to progress in terms of WSH management.

After the contractor has been selected, there must be frequent meetings to monitor how the contractor is progressing in its work and whether critical WSH procedures are implemented, e.g. risk assessment, inspections, WSH meetings, incident investigation and audits. The contractors' resources such as plants, machinery, equipment and chemicals must be checked for their compliance to the company's requirements before they can be brought into the workplace. There can also be the sharing of incident learning and good practices among the contractors or sub-contractors.

Outsourcing has similarities to the engagement of a contractor. The key difference is that the function or process being outsourced can actually be performed in-house, but the organisation chose to focus on their core business and outsource non-core functions or processes. The key reasons for outsourcing are usually cost, resources and lack of time. As in the case of engaging contractors, the outsourced companies must be carefully selected and monitored.

Organisations also need to identify and prepare for the emergencies that can occur in their workplaces. The processes to prepare for and respond to these emergencies need to be established, implemented and maintained. These processes include establishing an emergency response plan (including first aid), training for relevant personnel, periodic tests and exercises, performance evaluations, and communication of the plan to workers and interested parties.

6.3.8 Performance evaluation (Clause 9)

The key elements under performance evaluation include monitoring, measurement, analysis and performance evaluation (clause 9.1), and internal audit (clause 9.2). The performance evaluation processes cover both quantitative and qualitative measures of WSH performance. These will be closely related to the policies, objectives and targets. At the same time, the measures will also be guided by the risk assessment, especially the high-risk hazards and critical controls. The measures can also be split into proactive and reactive measures. Proactive measures monitor the implementation of controls and management activities, which are not dependent on the occurrence of incidents and accidents, i.e. reactive measures. Some of the proactive measures include inspections, audits, evaluation of training effectiveness, behaviour-based observations, and perception or safety culture/climate surveys. Reactive measures count the number and rate of incidents and accidents, e.g. accident frequency rate, accident severity rate and fatality rate.

Evaluation of compliance is directly related to the legal and other WSH requirements identified during planning. During implementation and operations, there should be periodic audits and assessments to determine if the organisation is complying with all applicable legislations and standards.

An audit is an important activity in the evaluation of compliance to the management system. This is unlike inspection, which is usually only focused on event level issues. According to ISO 45001:2018, an audit is a "systematic, independent and documented process for obtaining 'audit evidence' and evaluating it objectively to determine the extent to which 'audit criteria' are fulfilled". There is a strong emphasis on evidence and the evaluation needs to be thorough. An audit is usually based on the review of documents, interviews with employees and physical inspections. The three processes provide triangulation so that the auditor can form a credible impression of how well the management system is established and maintained. The audits can be conducted internally or by an external auditor. In the Singapore construction industry, for example, the Building and Construction Authority (BCA) and the Ministry of Manpower (MOM) both require safety and health audits on construction contractors. The BCA Requires OHSAS 18001 (predecessor of ISO 45001:2018) /SS506 and ISO 14000 and ISO 9000 certification for contractors with a rating of B2 and above. Certification audits are usually conducted by independent and external auditors. These contractors need to be audited annually. In contrast, MOM requires six monthly independent and external audits for projects that are S\$30 million and above, while projects with a contract sum of less than S\$30 million are required to conduct internal reviews of the management system every six months. Many accidents have shown that audits can be compromised due to commercial pressure on the auditors. Auditors can be pressurised by their paymaster, the contractors, to be more lenient. Thus, the professionalism of the auditors and the WSH culture of the contractors play an important role in the success of the audit regimes.

"Management review" refers to a high-level review of how effectively the management system is performing, which allows top managers to make necessary changes to system elements like policies, resources and processes, to improve the performance. There is a wide range of documents that the management can review to get a sense of how established, implemented and well-maintained the WSH management system is. These documents would include the incident investigation reports, audit reports and performance indicators derived from objectives and targets.

6.3.9 Improvement (Clause 10)

Incident investigation is one of the most fundamental management processes and the organisation must respond strongly and seriously to any incident that occurs. This is covered in more detail in Chapter 3.

Besides incidents, nonconformity can be uncovered during operations or inspections, such that employees and contractors may be found to violate WSH rules or procedures. Based on investigations into the incidents or nonconformity, corrective actions will be proposed, and these should be recorded and tracked for their effectiveness and implementation. Corrective actions are focused on removing causes of nonconformity and preventing their recurrence. On the other hand, opportunities for continual improvement can also be identified during the investigation or operations. These opportunities are not linked to any specific nonconformity or incidents, but they do enhance WSH performance.

6.4 Conclusions

The workplace safety and health (WSH) management system manages the risk of WSH incidents. It is important for managers to think systemically and design processes and structures that ensure that suitable and effective controls are identified, implemented and maintained. This will then prevent the unsafe acts and conditions that cause incidents to happen. ISO 45001:2018 is a comprehensive document that provides a useful illustration of the key components expected of a robust and reliable WSH management system.

Review Questions

1. What is the difference between a control measure and an element or sub-system of a management system?
2. Give three examples of WSH Policy, Objectives and Targets for a construction contractor specialising in excavation work.
3. Explain how contractors and suppliers should be managed in terms of WSH.
4. What are the audits required by BCA and MOM?
5. Explain what management review is and why it is important.

References

American National Standard (ANSI)/ American Society of Safety Engineers (ASSE). (2017). *ANSI/ASSE Z10-2012 (R2017) Occupational Health and Safety Management Systems*.

BSI Standards Limited. (2018). *BS 45002-0:2018 Occupational health and safety management systems — Part 0: General guidelines for the application of ISO 45001*.

DNVGL. (n.d.). ISRS™: for the health of your business — Oil & Gas — DNV GL. Retrieved April 3, 2018, from https://www.dnvgl.com/services/isrs-for-the-health-of-your-business-2458

ExxonMobil. (n.d.). OIMS: A disciplined management framework | Exxon-Mobil. Retrieved April 3, 2018, from http://corporate.exxonmobil.com/en/company/about-us/safety-and-health/operations-integrity-management-system

International Labor Organization (ILO). (2001). Guidelines on occupational safety and health management systems . Retrieved from http://www.ilo.org/wcmsp5/groups/public/---dgreports/---dcomm/---publ/documents/publication/wcms_publ_9221116344_en.pdf

ISO. (2018). *ISO 45001:2018 Occupational health and safety management systems*.

Singapore Standards Council. (1999). *Code of practice for safety management system for construction worksites*.

7 Safety Culture and Leadership

The Event Causation Technique

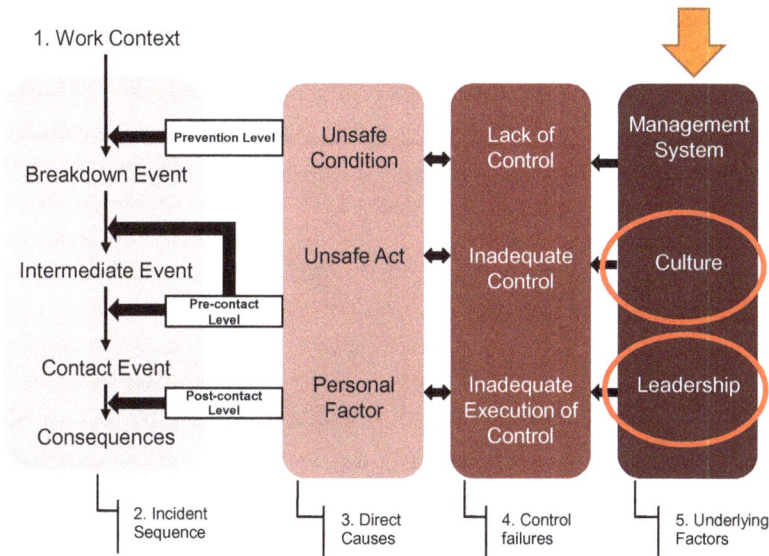

1. Work Context

Breakdown Event — Prevention Level

Intermediate Event — Pre-contact Level

Contact Event — Post-contact Level

Consequences

Unsafe Condition ⟷ Lack of Control ← Management System

Unsafe Act ⟷ Inadequate Control ← Culture

Personal Factor ⟷ Inadequate Execution of Control ← Leadership

2. Incident Sequence | 3. Direct Causes | 4. Control failures | 5. Underlying Factors

Safety culture and leadership are underlying factors within the Event Causation Technique (ECT) framework (see ECT diagram above). They influence the effectiveness of the management system and the risk controls. Unlike management systems, culture and leadership are softer aspects of workplace safety and health (WSH) management which may not be easy to evaluate and improve. Nevertheless, they are critical components of WSH management that have wide-ranging impact.

7.1 Introduction

Culture affects all management processes in an organisation because it is defined as the shared values and beliefs of the people in the organisation. A procedure or risk assessment may be very comprehensively written, but these detailed analyses or instructions can be easily defeated if the workers on the ground and the people in supervisory or management positions are not committed to them. In addition, leaders are the key people who can influence safety culture. This chapter will first discuss the definition of safety culture by introducing several safety culture models. Subsequently, safety leadership will be briefly introduced. Finally, the BP Texas City plant explosion case will be discussed to illustrate the importance of safety culture and leadership.

7.2 Defining Safety Culture

To define safety culture, we need to first understand organisational culture. One widely used definition of organisational culture is, "a system of shared values (*what is important*) and beliefs (*how things work*) that interact with a company's people, organisational structures, and control systems to produce behavioural norms (*the way we do things around here*)" (Uttal, 1983). Culture can be very abstract and not clearly defined because it is made up of values and beliefs which are embedded in the minds of the people in the organisation. Culture can be observed through the pattern of behaviours across people, time and space.

Using the analogy of weather and climate, localised behaviours or specific events are like weather, which can vary in the short term. However, climatic patterns are relatively stable across longer spans of time and space. For instance, December is usually a very rainy month in Singapore, but it does not mean that there is rain every day in December. The longer-term pattern of December being a wet month is similar to organisational culture, which is relatively stable across the years, but there is occasional deviation from the general pattern, meaning that individuals may still exhibit values, beliefs and behaviours not aligned with the organisation's culture from time to time.

Following the same analogy, although the weather in Singapore is relatively uniform across the small island, the Meteorological Service Singapore has indicated that "rainfall is higher over the northern and western parts of Singapore and decreases towards the eastern part of the island". Likewise, organisational culture is not always uniform, even in small organisations. This is because each individual's values, beliefs and behaviours are always influenced by other factors such as the situation the person is in, the individual's experience and upbringing, personality, group influences, etc.

Applying the definition of organisational culture to safety, safety culture can be defined as a system of shared values and beliefs that interact with a company's people, organisational structures, and control systems to produce *safety-related* behavioural norms (see Figure 7.1). The behaviour of each individual person in the organisation is affected by the individual's characteristics, the safety culture (shared values and beliefs) of the organisation and the management system (structure and control system). Since all individuals are similarly influenced by the safety culture and management system, there will be similarities in the behaviours

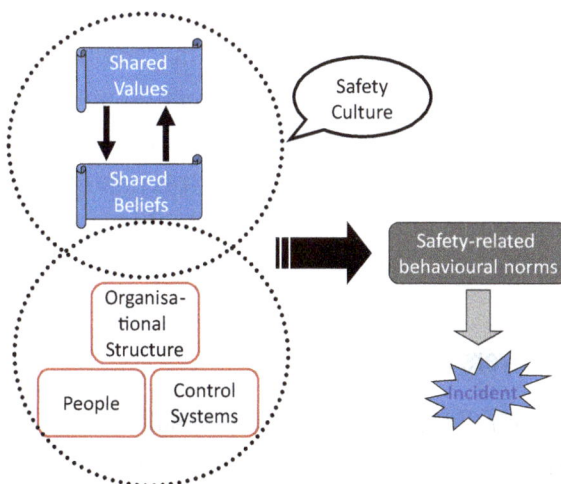

Figure 7.1   Safety culture

of all individuals in the organisation, but some deviation is expected due to differences in individual characteristics.

This definition of safety culture is aligned with the Event Causation Technique (ECT), which highlights that unsafe behaviour (a direct cause of incidents) is influenced by risk controls and management system components and safety culture. This concept is also highlighted by numerous researchers, for example Guldenmund (2007), who indicated that individuals' behaviours are influenced by the organisation's structure, culture and processes, which are "dynamically interrelated". He also indicated that an "organisation's culture cannot be isolated from its structure or processes".

To understand safety culture further, it is important to define beliefs and values. Beliefs refer to basic assumptions that a group of people hold in their minds. Some examples are, "safety is costly", "we have no time for safety and health issues", "accidents are a matter of luck" and "it all boils down to workers' behaviours; there is not much that the management can do". Beliefs drive behaviours. Hence, if one has the wrong beliefs, their decision-making and intentions can be unsafe. Values refer to the relative importance of different concepts, e.g. safety versus productivity and integrity versus ambition. Shared values and shared beliefs are the common values and beliefs in a group of people. The common values and beliefs arise or were inculcated through shared experiences, stories and knowledge. The management system or norms in the group will help to create and maintain the culture in a workplace.

As culture is intangible, many managers do not focus on culture when they are managing the organisation. This can lead to management actions that fail to deliver or even backfire. It is important to design the management system to be aligned with the desired culture, while taking the current culture into consideration.

7.3 Different Layers of Culture

Based on previous research in organisational culture, Guldenmund (2000) presented a multi-level model of safety culture (see Figure 7.2).

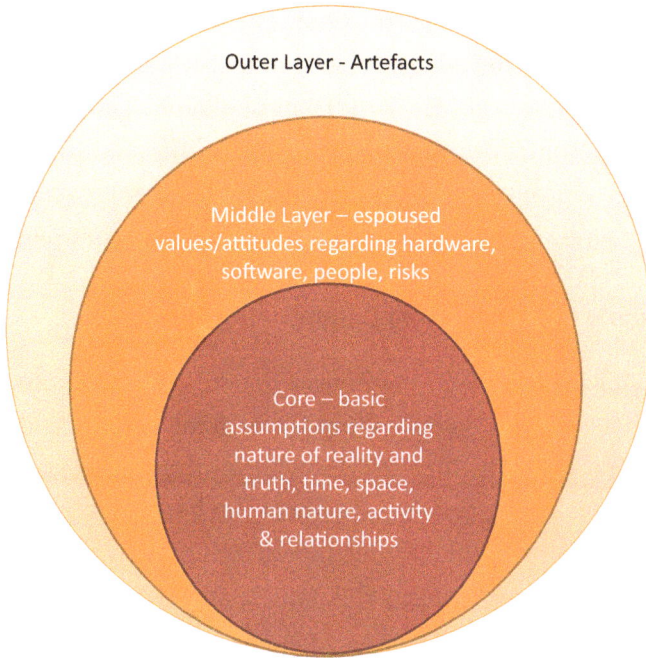

Figure 7.2   Levels of culture

The core of safety culture is assumed to comprise shared basic assumptions, which are the shared beliefs presented earlier. These assumptions include the nature of reality and truth, time, space, human nature, human activity and human relationships. They are mainly implicit, but obvious or even factual to members of the organisation, yet they remain implicit, invisible and pre-conscious (i.e. assumed to be true in an unconscious fashion). These basic assumptions, or shared beliefs and values, have to be deduced from the middle and outer layers of culture.

The middle layer is made up of espoused *values or attitudes* regarding hardware, software, people and risks. This layer is more explicit and conscious, and it includes people's attitudes towards objects or artefacts such as organisational policies, training manuals, procedures, formal statements, bulletins, accident and incident reports, job descriptions and minutes of meetings. Finally, the outer layer refers to the objects or artefacts that are visible and representative of safety culture, such as those mentioned earlier. Other examples of artefacts include management statements, meetings, inspection reports, dress codes, personal protective equipment (PPE), posters and bulletins. However, artefacts are poor representations of the underlying culture and they are easily misinterpreted. For example, a company with many safety posters is not necessarily serious about the safety and health of its employees.

7.4 Reason's Model of Safety Culture

Chapter 9 of Reason (1997) presents a model of safety culture which assumes that a positive safety culture is essentially an informed culture. An informed culture is, "one in which those who manage and operate the system have current knowledge about the human, technical, organisational and environmental factors that determine the safety of the system as a whole." A truly informed culture is in a sustained state of intelligent and respectful wariness. The organisation gathers the right kind of data using a safety information system. They do not just rely on lagging indicators; they also use leading indicators to help them manage safety, health and environmental risks proactively.

Reason (1997) felt that a safety culture can be socially engineered by implementing the different components of an informed culture. To him, an informed culture is based on four sub-cultures: reporting culture, just culture, flexible culture and learning culture. All four sub-cultures interact and come together to produce an organisation that has an informed culture.

7.4.1 Reporting culture

Reporting culture refers to the involvement and willingness of people to report hazards, errors and incidents. Possible disincentives to reporting culture include concerns about the impact of reporting, perceived additional work or trouble. To encourage reporting, there could be indemnity against disciplinary proceedings (as far as practicable), and ensuring confidentiality or de-identification of reporting staff.

There must be an independent and credible department or agency for disciplinary actions, which is separate from the department receiving reports of hazards and making WSH improvements. Other incentives include prompt feedback to the community about the actions and assessment of reported hazards or incidents and ensuring the ease of making the reports. The aviation industry has implemented successful reporting programmes by implementing many of these incentives for reporting. The success factor for a reporting system is trust in the system. Reporters must understand the importance of reporting and feel safe in reporting even their own errors.

7.4.2 Just culture

Just culture is essentially having a clearly established set of safety rules, responsibilities and accountabilities such that everyone in the organisation knows what is expected of them and accept the consequences that come with violating the safety rules. Just culture is about the way that the organisation handles blame and punishment. A 'no-blame' culture is neither feasible nor desirable, as it may breed irresponsibility among some individuals and this can spread across the organisation. Conversely, a punish-no-matter-what approach is also not feasible.

Any punishment needs to take the circumstances of the unsafe act into consideration. Even if the unsafe act was intentional, the worker committing the unsafe act need not be culpable. Managers need to consider if the workers were encouraged to violate the rules due to lack of proper supervision or lack of effective training. It could also be a norm for everyone to break the rules, because of the management lack of interest in WSH management. Decisions on the punishment also need to take into consideration if the same unsafe act would be committed if another reasonable man was placed in the same situation. In addition, the safety history of the worker needs to be factored in. These questions need to be considered prior to any punishment.

Table 7.1 Effect of reward/punishment on behavioural change; adapted from Reason (1997)

	Immediate	**Delayed**
Reward	Positive effects	Doubtful effects
Punishment	Doubtful effects	Negative effects

It is important to note that just culture promotes an atmosphere of trust and helps to engender a reporting culture. There must be a good balance between punishment and rewards because rewards are known to produce positive effects in developing safe behaviour. With reference to Table 7.1, immediate recognition or other rewards are seen to be able to produce greater motivation to be safe. Some examples include giving workers small gifts or badges during site inspections as recognition of their positive safety behaviour. This is more effective in motivating safe behaviour than scolding and punishment.

7.4.3 Flexible culture

Flexible culture refers to the ability of an organisation to re-arrange its structure based on the needs of the work context. Flexible culture allows people with the right expertise to provide input or take the lead when their expertise is needed. This is especially prominent in crisis situations where the project manager or supervisors may not have the necessary expertise to manage the crisis. This may call for the fire safety manager or workplace safety and health manager to take over management of the site during the crisis. This can also happen during work, where those with different expertise are called upon to identify or control hazards. The ability to tap into expertise to suit the situation is a mark of flexible culture.

One of the key elements of flexible culture is the knowledge of where different expertise lie in the organisation. It is important for the safety information system to contain a comprehensive database of the strengths and competencies of different personnel within the organisation. However, members of the organisation must also be willing to offer their expertise readily. Proactive involvement can be encouraged through incentives, but a sense of teamwork is more fundamental to a flexible culture.

7.4.4 Learning culture

Learning culture is the ability to learn from past incidents and available information to prevent future incidents. This involves detailed analysis of information and having the ability to learn as a group. The ability to look beyond events and spot trends and patterns so as to prevent undesirable events is a key mark of a learning organisation. Learning culture requires an open mind and the willingness to listen to others.

However, an organisation need not have accidents to identify ways to improve their WSH management. The company can actively search for opportunities to learn by, for example, evaluating incidents from companies with similar operations, collecting feedback from workers periodically, reviewing audit reports, and analysing complaints from customers or other stakeholders. Another useful approach is to invite sister units or peer organisations to conduct benchmarking or peer review exercises to help the organisation review its performance and adopt best practices. It is also important to learn from successful departments or individuals within organisations and understand their key success factors. Improvement teams can be set up to innovate and focus on specific WSH issues. These improvement teams can be recognised and incentivised through competitions and awards.

7.5 HSE Safety Culture Questionnaire

Safety culture is intangible and hard to manage. Thus, many researchers and organisations have developed questionnaires to help organisations assess safety culture. Many researchers use the term safety climate instead of safety culture because safety climate refers to the middle layer of safety culture (see Figure 7.2), i.e. attitudes, while safety culture refers to the core, which is pre-conscious and hard to measure. Many questionnaires use a quantitative approach, but some use a qualitative approach, combining responses from site personnel and the inspector or auditor so that safety culture improvements can be determined. One of these questionnaires is the HSE safety culture questionnaire (Health and Safety Executive, n.d.), which is made up of seven key factors:

1. Management commitment
2. Communication
3. Employee involvement
4. Training/information
5. Motivation
6. Compliance with procedures
7. Learning organisation

In contrast to the four sub-cultures of Reason's model, the HSE safety culture questionnaire highlights the importance of management commitment and the need for managers to be seen leading by example as far as safety is concerned. Another important aspect of management commitment is how managers allocate resources and how they make tough decisions when safety and productivity or profit contradict each other. A good example of strong management commitment to safety is how Paul O'Neill, CEO of Alcoa (1987–2000) turned around the safety records of the company, and how he, while maintaining these impressive safety records, made Alcoa one of the most profitable companies in the world (Duhigg, 2014; Roth, 2012).

Learning organisation is synonymous with learning culture in Reason's model. Communication, employee involvement, information and motivation are related to reporting culture and just culture in Reason's model. The HSE safety culture questionnaire also highlights the importance of training and compliance, which are not highlighted in Reason's model. However, some of these components of safety culture are closely

Organization-Level Safety Climate

Top management in this plant–company...

1. Reacts quickly to solve the problem when told about safety hazards.
2. Insists on thorough and regular safety audits and inspections.
3. Tries to continually improve safety levels in each department.
4. Provides all the equipment needed to do the job safely.
5. Is strict about working safely when work falls behind schedule.
6. Quickly corrects any safety hazard (even if it is costly).
7. Provides detailed safety reports to workers (e.g., injuries, near accidents).
8. Considers a person's safety behaviour when moving/promoting people.
9. Requires each manager to help improve safety in his/her department.
10. Invests a lot of time and money in safety training for workers.
11. Uses any available information to improve existing safety rules.
12. Listens carefully to workers' ideas about improving safety.
13. Considers safety when setting production speed and schedules.
14. Provides workers with a lot of information on safety issues.
15. Regularly holds safety-awareness events (e.g., presentations, ceremonies).
16. Gives safety personnel the power they need to do their job.

Note. Items cover three content themes: Active Practices (Monitoring, Enforcing), Proactive Practices (Promoting Learning, Development), and Declarative Practices (Declaring, Informing).

Group-Level Safety Climate

My direct supervisor...

1. Makes sure we receive all the equipment needed to do the job safely.
2. Frequently checks to see if we are all obeying the safety rules.
3. Discusses how to improve safety with us.
4. Uses explanations (not just compliance) to get us to act safely.
5. Emphasizes safety procedures when we are working under pressure.
6. Frequently tells us about the hazards in our work.
7. Refuses to ignore safety rules when work falls behind schedule.
8. Is strict about working safely when we are tired or stressed.
9. Reminds workers who need reminders to work safely.
10. Makes sure we follow *all* the safety rules (not just the most important ones).
11. Insists that we obey safety rules when fixing equipment or machines.
12. Says a "good word" to workers who pay special attention to safety.
13. Is strict about safety at the end of the shift, when we want to go home.
14. Spends time helping us learn to see problems *before* they arise.
15. Frequently talks about safety issues throughout the work week.
16. Insists we wear our protective equipment even if it is uncomfortable.

Note. Items cover three content themes: Active Practices (Monitoring, Controlling), Proactive Practices (Instructing, Guiding), and Declarative Practices (Declaring, Informing).

Figure 7.3 Questionnaire items developed by Zohar and Luria (2005)

related to segments of the WSH management system components. This is not surprising, because safety culture affects the way the WSH management system is developed, implemented and maintained. A well-implemented WSH management system is reflective of a positive safety culture.

7.6 Zohar's Safety Climate Survey

Zohar (Hofmann *et al.*, 2017; Zohar, 1980; Zohar, 2010; Zohar and Luria, 2005) is known for his research on safety climate. Figure 7.3 shows an example of the questionnaire items in a typical safety climate questionnaire. Each statement will be rated by the respondent on a Likert scale of 1 to 5, or 1 to 7. The aggregated score will then give the management a sense of the safety climate of the organisation. In addition, it is possible to target specific management or supervisor behaviours to improve the safety climate. Zohar (2010) opined that safety climate is "a robust leading indicator or predictor of safety outcomes across industries and countries." The periodic measurement of an organisation's safety climate provides the organisation with another leading indicator, which must be read in conjunction with other leading and lagging indicators.

7.7 CultureSAFE and Others

The Singapore's Workplace Safety and Health Council (2015) has a comprehensive safety culture assessment and improvement programme called CultureSAFE. CultureSAFE provides the safety culture diagnostic tool, which is a questionnaire survey, and a range of recommended programmes to improve safety culture. Another well-known safety culture improvement programme is the Hearts and Minds programme (Energy Institute n.d.), developed by Shell. The range of safety culture improvement programmes is very wide and each has a different take on what safety culture entails. Nevertheless, the different models agree that safety culture is about human and organisational aspects of Safety, Health and Environmental (SHE) management. Safety culture influences how the SHE management system and risk controls are developed, implemented and maintained.

7.8 Safety Leadership

The range of leadership models is very wide. This section will briefly discuss transformational leadership and transactional leadership in the context of WSH. One of the key leadership approaches adopted by many leaders to influence subordinates' safety behaviour is the use of reward and punishment systems. This is known as transactional leadership (Northouse, 2016). In contrast, transformational leadership "evokes changes in subordinates' value systems to align them with organizational goals" (Clarke, 2013) and numerous studies have shown that it has positive effects on the safety behaviour and safety participation of subordinates. This section will highlight how leaders can adopt the transformational leadership style so as to engender positive safety culture in their workplace.

7.8.1 Transformational leadership

Many leadership theories were developed over the years, but since Downton (1973) coined the term *Transformational Leadership*, it has become one of the most widely researched leadership theories (Northouse, 2016). So what is *Transformational Leadership*? It is widely accepted that transformational leadership encompasses the following distinct but correlated dimensions (aka the 4 'T's):

1. Idealised influence, i.e. leader instils confidence and behaves in admirable ways that cause the followers to identify with him/her
2. Inspirational motivation, i.e. leader inspires others towards goals, provides meaning, optimism and enthusiasm, and articulates a vision that is appealing and inspiring to others
3. Intellectual stimulation, i.e. leader challenges assumptions, takes risks and encourages subordinates to be creative

4. Individualised consideration, i.e. leader shows interest in subordinates' personal and professional development and listens to followers' needs and concerns

7.8.2 Transformational leadership and safety culture

At the same time, much research has shown that WSH leadership is one of the key criteria of a positive safety climate or culture (Zohar, 2010). In essence, "leaders create climate" (Lewin *et al.*, 1939). To facilitate ease of discussion, climate is deemed to be a proxy to culture in this section, i.e. they are treated as "pretty much the same thing". In the case of transformational leadership, it is quite intuitive that transformative leadership will have a positive impact on safety culture. A closer look at each of the 'I's in transformational leadership will shed more light on the relationship between transformational leadership and safety culture (adapted from Clarke (2013)):

1. Idealised Influence: managers adopting transformational leadership are role models by doing what is morally right as compared to focusing on production goals only. Idealised influence encourages focus on WSH and sustainable ways of working, in contrast to prioritising short-term benefits due to work pressure. Leaders high in idealised influence will demonstrate their personal commitment to WSH as a core value. The personal commitment will improve followers' trust in management and loyalty, which can lead to improvement in overall performance.
2. Inspirational motivation: transformational leaders motivate followers to be part of the organisation's shared vision and to go beyond their individual needs. In terms of WSH, transformational leaders inspire their followers to achieve safety levels previously believed to be impossible, e.g. Vision Zero. Transformational leaders frequently use symbols and stories to articulate their vision and inspire followers.
3. Intellectual stimulation: Intellectual stimulation arises when leaders encourage followers to address WSH issues and enhance information sharing by challenging long-held assumptions and promoting

innovation in WSH. This engenders a learning culture, where followers becomes more aware of and better understand WSH problems, encouraging innovative, rather than reckless solutions.

4. Individualised consideration: transformational leaders have a strong and sincere interest in their followers' well-being, which naturally includes WSH. Transformational leaders are not satisfied with compliance with WSH legislations because they want to ensure that their followers do not experience injuries and ill-health.

7.8.3 Empirical evidence

The positive impact of transformational leadership on safety culture and safety performance is supported by empirical studies. For example, Barling *et al.* (2002) studied 174 restaurant workers and 164 young workers from diverse jobs using selected questions from the Multifactor Leadership Questionnaire (MLQ) (Bass and Avolio, 1990), a shortened version of the safety climate questionnaire by Zohar (1980), and they also measured the self-reported safety consciousness and safety incidents. The study showed that, among other factors, safety-specific transformational leadership plays a significant role in influencing "safety climate, safety consciousness, and safety-related events." On the other hand, passive leadership is shown to be related to poorer safety climate (Kelloway *et al.*, 2006). Thus, past studies have shown that leadership influences climate, and in turn, climate had been shown to influence occurrence of workplace injuries (Huang *et al.*, 2012). Transformational leadership is a potentially powerful tool to influence safety culture and WSH.

7.8.4 Transformational leadership practices

Based on the 4 'I's, leaders can adopt the following practices to be more aligned with transformational leadership (Kouzes and Posner, 2002):

1. Model the way: Be clear about own values and philosophy and be a role model.

2. Inspire a shared vision: Develop a compelling vision that helps followers visualise positive outcomes and see how their dreams can materialise.
3. Challenge the process: Transformational leaders are pioneers because they are willing to experiment, try new things and learn from mistakes so as to achieve better results.
4. Enable others to act: Outstanding leaders are effective at working with people by building trust, promoting collaboration and listening respectfully. They empower others to make decisions and feel good about their work by linking it with greater purposes.
5. Encourage the heart: Be attentive to followers' need for recognition and reward. Show appreciation and encouragement to others.

7.8.5 Discussion

Transformational leadership has the potential to improve the performance of organisations, including safety culture and hence WSH performance. Positive changes in followers' behaviour will take time, but the persistent implementation of transformational leadership will reap results in WSH and organisational performance in general. This is not saying that it is the only leadership style that is applicable to all situations. Successful leaders will have the wisdom to observe and apply suitable leadership actions in different situations so as to improve WSH. In the end, leaders have to take it upon themselves to use transformational leadership or other approaches to convince their followers to adopt safer work methods.

7.9 BP Texas Explosion Case Study

The BP Texas City Refinery explosion and fire (Chemical Safety Board, 2007) is one of the worst industrial accidents in recent history. The accident happened on 23 March 2005 at 1:20pm (GMT-5). The explosion and fire killed 15 people and injured another 180. The financial losses were estimated to be at least US$1.5 billion. The explosion occurred during the start-up of an isomerisation (ISOM) unit, when a raffinate splitter tower was overfilled. The overfilling led to the opening of pressure relief devices (a safety measure), but it resulted in a flammable liquid geyser (a fountain of high pressure jets of fluid and gas that shoots into the air) from a blowdown stack that did not have a flare (a safety device which can burn gas or liquid released through the blowdown).

The investigations conducted after the accident revealed severe organisational and safety culture problems in BP Texas City Refinery. Some of them are provided below:

1. The BP Board of Directors did not provide effective oversight of BP's safety culture and major accident prevention programmes. The Board did not have a member responsible for assessing and verifying the performance of BP's major accident hazard prevention programmes. *Note: This shows that safety leadership was inadequate. An informed culture where senior management were in the know of safety issues was not present.*
2. Numerous surveys, studies, and audits identified deep-seated safety problems at Texas City, but the response of BP managers at all levels was typically "too little, too late." *Note: This is a sign of lack of proactive actions by managers and leaders, i.e. safety leadership issue.*
3. Cost-cutting, failure to invest and production pressures from BP Group executive managers impaired process safety performance at Texas City. *Note: This reflected how cost-cutting was valued over process safety; a sign of poor safety culture and a lack of safety leadership.*
4. A "check the box" mentality was prevalent at Texas City, where personnel completed paperwork and checked off on safety policy and procedural requirements even when those requirements had not been met. *Note: Routine violation was common and such behaviour is again a reflection of unsafe shared values and beliefs.*
5. BP Texas City lacked a reporting and learning culture. Personnel were not encouraged to report safety problems and some feared retaliation

for doing so. The lessons from incidents and near-misses, therefore, were generally not captured nor acted upon. Important relevant safety lessons from a British government investigation of incidents at BP's Grangemouth, Scotland refinery were also not incorporated at Texas City. *Note: This is directly aligned with Reason's (1997) model of safety culture, where reporting and learning culture are critical components.*

6. Reliance on the low personal injury rate at Texas City as a safety indicator failed to provide a true picture of the processes' safety performance and health of the safety culture. *Note: This is a poor choice of WSH indicator, is a management system issue.*

7. Safety campaigns, goals, and rewards focused on improving personal safety metrics and worker behaviours rather than on process safety and management safety systems. While compliance with many safety policies and procedures was deficient at all levels of the refinery, Texas City managers did not lead by example regarding safety. *Note: Similar to the above point, this is a major management system issue.*

The above investigation findings show that inadequate WSH management, poor safety culture and inadequate safety leadership were major contributors to the accident. These underlying factors are almost always found in major accidents, albeit with have some variance in the details.

7.10 Conclusions

Safety culture is about an organisation's shared values and beliefs that influence safety related behaviours. There are various models of safety culture and only a few were discussed in this chapter. Regardless of the definitions and models, the key thing to note is that safety culture influences, albeit not in a deterministic manner, human behaviours while implementing workplace safety and health (WSH) procedures and policies. If there is no buy-in from the people implementing the

WSH management system, the system will never work. When dealing with people, we need to use humanistic language and be in sync with emotions. Despite the importance of legislations, liabilities and statistics, if there is an over-focus on them, it is not possible to get WSH messages across. Leaders and managers have to demonstrate genuine care for the workers to let the workforce know that everybody needs to use and improve the WSH management system. Thus, safety leadership is intertwined with safety culture and the keys to a positive culture are the leaders.

Review Questions

1. Explain Reason's model of safety culture and contrast it with the HSE Safety Culture questionnaire.
2. Why is top management so important in improving safety culture?
3. What can top managers do to demonstrate their commitment to WSH?
4. Differentiate transactional and transformational leadership.
5. List some of the safety culture problems at BP Texas City Refinery prior to the accident in 2005.

References

Barling, J., Loughlin, C., and Kelloway, E. K. (2002). "Development and test of a model linking safety-specific transformational leadership and occupational safety." *Journal of applied psychology*, 87(3), 488.

Bass, B. M., and Avolio, B. J. (1990). *Transformational leadership development: Manual for the multifactor leadership questionnaire*, Consulting Psychologists Press.

Chemical Safety Board (2007). "Investigation Report — Refinery Explosion and Fire (BP Texas)." <http://www.csb.gov/>. (December, 2008).

Clarke, S. (2013). "Safety leadership: A meta-analytic review of transformational and transactional leadership styles as antecedents of safety behaviours." *Journal of Occupational and Organizational Psychology*, 86(1), 22–49.

Downton, J. V. (1973). *Rebel leadership: Commitment and charisma in the revolutionary process*, Free Press.

Duhigg, C. (2014). *The power of habit: Why we do what we do in life and business*, Random House, New York.

Energy Institute (n.d.). "Hearts and Minds." <http://www.eimicrosites.org/heartsandminds/>. (Oct 8, 2015).

Guldenmund, F. W. (2000). "The nature of safety culture: A review of theory and research." *Safety Science*, 34(1–3), 215–257.

Guldenmund, F. W. (2007). "The use of questionnaires in safety culture research — an evaluation." *Safety Science*, 45(6), 723–743.

Health and Safety Executive (n.d.). "Common topic 4: Safety culture." <www.hse.gov.uk/humanfactors/topics/common4.pdf>.

Hofmann, D. A., Burke, M. J., and Zohar, D. (2017). "100 Years of occupational safety research: From basic protections and work analysis to a multilevel view of workplace safety and risk." *Journal of Applied Psychology*, 102(3), 375–388.

Huang, Y.-H., Verma, S. K., Chang, W.-R., Courtney, T. K., Lombardi, D. A., Brennan, M. J., and Perry, M. J. (2012). "Supervisor vs. employee safety perceptions and association with future injury in US limited-service restaurant workers." *Accident Analysis & Prevention*, 47, 45–51.

Kelloway, E. K., Mullen, J., and Francis, L. (2006). "Divergent effects of transformational and passive leadership on employee safety." *Journal of occupational health psychology*, 11(1), 76.

Kouzes, J. M., and Posner, B. Z. (2002). *The leadership challenge*, Jossey-Bass, San Francisco.

Lewin, K., Lippitt, R., and White, R. K. (1939). "Patterns of aggressive behavior in experimentally created "social climates." *The Journal of social psychology*, 10(2), 269–299.

Northouse, P. G. (2016). *Leadership: theory and practice*, SAGE, Thousand Oaks.

Reason, J. (1997). *Managing the risks of organizational accidents*, Ashgate, Aldershot.

Roth, M. (2012). "'Habitual excellence': The workplace according to Paul O'Neill." <http://www.post-gazette.com/business/businessnews/2012/05/13/Habitual-excellence-The-workplace-according-to-Paul-O-Neill/stories/201205130249>. (Oct 8, 2015).

Uttal, B. (1983). "The corporate culture vultures." *Fortune*, Oct. 17, 66–72.

Workplace Safety and Health Council (2015). "Apply for CultureSAFE." <http://www.wshc.gov.sg/>. (Oct 8, 2015).

Zohar, D. (1980). "Safety climate in industrial organizations: Theoretical and applied implications." *Journal of Applied Psychology*, 65(1), 96–102.

Zohar, D. (2010). "Thirty years of safety climate research: Reflections and future directions." *Accid. Anal. Prev.*, 42(5), 1517–1522.

Zohar, D., and Luria, G. (2005). "A multilevel model of safety climate: Cross-level relationships between organization and group-level climates." *Journal of Applied Psychology*, 90(4), 616–628.

Improving Safety Culture

The Event Causation Technique

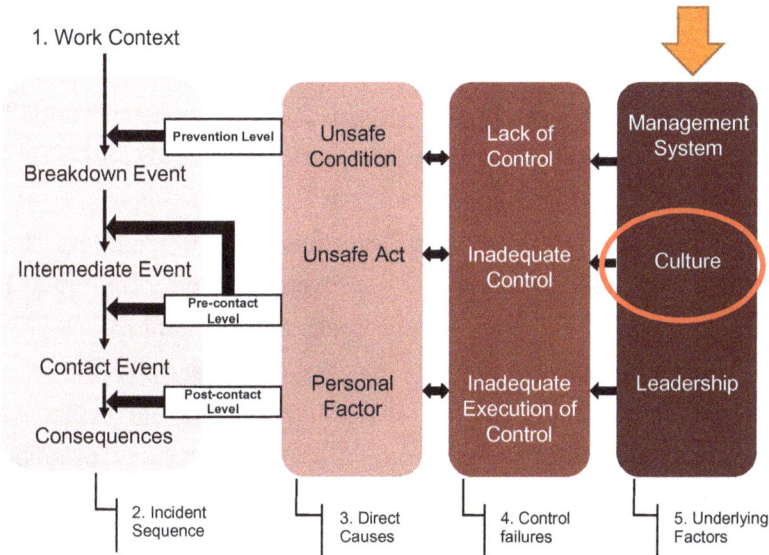

As established in the earlier chapters, safety culture interacts with management systems to influence behavioural norms and hence incident occurrence. To improve safety culture, the organisation need to undergo fundamental changes, which can be implemented based on organisational change theories and concepts.

8.1 Introduction

Organisational change is defined as a change to the *core* of an organisation. The *core* of an organisation includes organisational culture, organisational problem-solving capabilities, ability to cope with changes in the environment, organisational effectiveness, and congruence between organisational dimensions like environment, policy, structure and processes. This chapter provides a brief overview of how organisational change management concepts can be used to improve safety culture. Managers, project managers, engineers, WSH managers, and WSH officers will need to lead safety culture initiatives, which can constitute a significant organisational change.

In the context of workplace health and safety (WSH), change management usually refers to management of change (MOC), which is an operational level change that is systematic, risk-based, and aimed at controlling changes that can introduce hazards and cause accidents or ill-health. Instead of MOC, this chapter will be discussing change at the organisational level, specifically, improvement in safety culture. In the context of WSH, the motivation for improvement in safety culture is frequently a reaction to external pressures, e.g. enforcement from a regulator or client organisations after a major accident. Other possible sources of organisational change include leadership change, mergers, technological changes, growth and expansion of the company, new product(s), new competitors, legislative changes, political changes and changes in the demographics of customers.

After a major accident, an organisation's sustained success or even survival will depend on its ability to improve its safety culture and WSH management system in response to internal and external pressures. If an organisation is not able to demonstrate its ability to ensure workers' safety and health, they may be put out of business and leaders may be removed from their positions. Thus, organisations need to proactively improve their safety culture and WSH management system, and not only after a major accident. However, most organisational change initiatives fail, so it is important for us to understand how to design and implement successful organisational change. In this chapter, the key focus will be on approaches to improve safety culture, which influences the effectiveness of the WSH management system and risk controls.

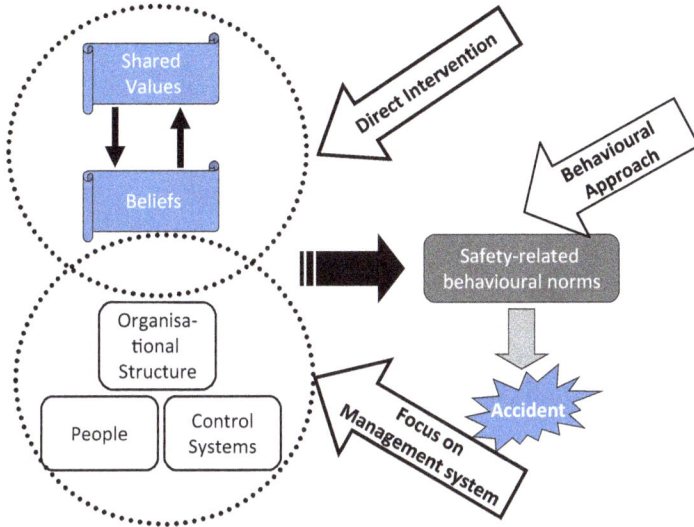

Figure 8.1 Approaches to improve safety culture

8.2 Improving Safety Culture

Using the definition of safety culture introduced in the last chapter, three possible approaches to improve safety culture can be identified (see Figure 8.1). First, a direct intervention approach can be taken. Direct intervention is focused on the middle layer of safety culture, which includes espoused values/attitudes such as expressing, communicating and diffusing desired safety beliefs and shared values throughout the organisation.

This can be done through facilitated focus group discussions, interviews and safety climate surveys. Focus group discussions and interviews will produce qualitative cases or stories on what employees see as anecdotal evidence of shared values and beliefs. These stories can be very powerful in further establishing shared values and beliefs. The effectiveness of directly discussing safety culture is highly dependent on the quality of communication and facilitation. A safety climate survey provides a quantitative source of information about the shared values and beliefs that can be used to triangulate along with the information obtained from focus group discussion and interviews. The direct intervention approach assumes that safety culture can be verbalised, discussed and modified through discussions.

Second, safety culture can be improved by focusing on the way the management system is established, implemented and maintained. This approach is an outside-in approach. Since management standards (e.g. ISO 45001 and ISO 14001) are essentially good practices adopted from organisations with positive safety culture, it is assumed that organisations that model themselves after these good practices will develop good safety culture. This is focused on the outer layer of safety culture, but since a management system includes training, communication and consultation, there will be overlaps with the direct intervention approach.

The last approach is a behavioural approach, which focuses on the behaviours of members of the organisations. Behaviours arise because of intention, motivation and consequences of the behaviours. They are more observable than values and beliefs, and are thus relatively good proxies for safety culture. A quantitative approach can then be taken in measuring behaviours. Tracking and trending the behaviours across time will provide managers with data which they can use as a leading indicator of safety management. During behavioural observations, the observer will be trained to provide positive reinforcement or feedback to motivate the person being observed to work safely. Such an approach is known as the behaviour-based safety (BBS) approach and there are many variations to how BBS is implemented. BBS assumes that external motivation will change behaviour in the long run, and as a result safety culture is improved.

However, many managers over-focus on the programmes for change. They neglect that change involves many factors beyond the programmes themselves.

8.3 Change Management Models

8.3.1 Lewin's change model

One of the well-known change models is Lewin's change model, which states that change goes from *unfreezing* to *moving* and then *refreezing*. *Unfreezing* involves the removal of resistance to change through understanding the reason for resistance and providing information and education to the relevant stakeholders. *Moving* refers to the process of changing behaviours, attitudes and values through the establishment of new policies, structure and processes. *Refreezing* creates a new equilibrium by reinforcing the changes that were made and ensuring their sustainability. Prior to the *unfreezing* stage, the organisation must identify the need for change and the problems that can arise if the change is not implemented. Throughout the change process, there is a need for data to be collected about the current situation, the change process, and the situation after the change process has been completed. In addition, the team overseeing the change will have to develop frequent communication and coordination opportunities for stakeholders to give feedback and share their challenges and concerns.

8.3.2 Rider-elephant-path model

Heath and Heath (2010) give an interesting analogy of the change process. According to them, each person has an emotional elephant and a rational rider. The rider riding the elephant is objective and rational, but may not be able to control the big elephant, which is focused on the short term and is guided by feelings. At the same time, if you want to change the behaviour of the rider and his elephant, you will have to shape the path that they are trudging on. Based on this analogy, Heath and Heath (2010) suggested nine key change management actions — three for each of the components in the analogy (the rider, the elephant and the path). Even though the following discusses each of the components separately, all three components interact with each other and it is difficult to clearly distinguish the effects of each component. In addition, not all nine change management actions are relevant in all situations. Change leaders need to select a suitable combination of actions to encourage change.

8.3.2.1 *Direct the rider*

The Rider is a "thinker and planner" but has a tendency to "over-think" (decision or analysis paralysis) and over-focus on problems. To direct the rider towards the desired behaviour, a leader should find "bright spots" (positive examples) for followers to follow, script the critical actions very clearly, and make the desired outcome crystal clear.

Finding the bright spot is about looking for "insiders" who have used good or best practices to achieve desired results or top performance. An important note is that the manager should look for *practice* and not just *knowledge*. For example, many contractors know that it is important to avoid work-at-height (knowledge), but very few actually *practise* work methods that minimise work-at-height consciously. Furthermore, it is important to emphasise bright spots that arose from the "locals" (of a project/company) so that the "not invented here" problem will not arise, and workers will be more receptive of the good practice. It is also interesting to note that many of the good or best practices are very small scale and simple. Once the bright spots are identified, leaders must thoroughly investigate why the bright spots are different from the rest. The success factors must be identified so that it is possible to clone as many of the bright spots as possible. In this way, the norm in the group changes. Cloning involves processes such as getting employees to shadow identified bright spots and getting bright spots to conduct sharing sessions.

Finding a bright spot is essentially a question of: "What's working and how can we do more of it?" It is solution-focused and not problem-focused. It is about scaling up solutions.

Since the rider can easily over-think, it is important for change leaders to narrow down the possible options, script the critical moves and direct the energy of workers towards the desired behaviour. Too many options delay change and employees may choose the *status quo* (i.e. refuse to change) if it is too difficult to decide. Thus, resistance to change may be due to lack of clarity on the desired behaviour. For individuals to implement the change, there must be crystal clear guidelines on the specific desired behaviour. Because it is not possible to script every single move, managers must especially script the critical moves. For example, a company might want workers to *be safe*, but what is "being safe"? This is where organisations need to define the specific safe behaviours for each type of trade and task.

Safe behaviour can be defined through safe work procedures (SWP), which were discussed in Chapter 4. Employers and principals are required to conduct risk assessments and based on the risk assessments, SWPs are derived. The SWPs have to document the specific risk controls that need to be implemented. Managers, supervisors and workers are expected to implement the risk controls, but if the SWPs were not clearly written and communicated, employees may not be able to implement the SWPs on the ground. The instructions for implementing the SWPs need to be written in a form that is easy for employees to understand.

Concurrently, organisations will need to make sure that the goals or desired outcomes are crystal clear so that the rider can focus his/her plans and decisions towards the goals. At the same time, having clear and desirable goals will also motivate the elephant.

8.3.2.2 *Motivate the elephant*

The elephant is the rider's biggest challenge. To motivate the elephant, find the feelings that influence people's behaviour (e.g. the joy of being with one's family), shrink the change into comfortable bite-sized or baby steps, and grow the people so that they feel a sense of identity and believe that they can change.

Heath and Heath (2010) cited Kotter and Cohen who observed that in most successful change efforts, the change process is not ANAL-YSE-THINK-CHANGE, but rather SEE-FEEL-CHANGE. In ANAL-YSE-THINK-CHANGE, a person analyses the situation, thinks of solutions and then changes to implement the solutions. The reality is that most of the time when we see the situation or something that grabs our attention, our emotions are stirred, and they influence our decisions and responses towards the situation. Thus, if the feeling that arises is not aligned with the intended change, change will not occur.

For example, if a campaign on "safe hands" is being implemented at a factory, even if many workers are invited to attend talks on how to protect their hands, because of the cognitive bias of over-confidence (discussed in Chapter 4), there is a tendency for the workers to feel that the likelihood of themselves being an accident victim are very small. Thus, the workers tend to ignore the safety measures promoted in such campaigns. The campaign did not motivate the elephant because the messages did not cause any emotion to arise. To motivate the elephant, it is more impactful to link the desired change or behaviours to positive emotions like happiness, camaraderie among workers, and recognition, rather than negative emotions like shame, fear of being punished and fear of losing one's family.

"Shrink[ing] the change" refers to cutting up a change into bite-size chunks so that it is not too daunting to implement. The elephant "hates doing things with no immediate payoff". Therefore, to evoke change, the first step or five-minutes are critical. If the employee is willing to start implementing the intended change, then there is a chance for the momentum to pick up so that the change might actually be implemented. The first task for effecting change should be simple and meaningful and employees should be recognised for their initial effort.

Another approach to motivate the elephant is to grow the people through inspiration, so that they become bigger than the problem. The crux is the identity that people adopt. If each employee sees himself or herself as a safety champion or the workplace as an extension of the family or friends, then helping to spot hazards and unsafe behaviour becomes natural. Thus, the key challenge for change leaders is how to make the intended change a matter of identity and not a cost-benefit analysis. The key test is this, "Would an employee (who is being asked to change) aspire to be the kind of person who would make this change?" If the answer is yes, the elephant can be more easily motivated to initiate the change. If the answer is no, then there is a need to review if the change is suitable or if the perceived identity needs to be re-designed. The change leader needs to be mindful of the following identity questions when considering their change initiative, "Who am I? What kind of situation is this? What would someone like me do in this situation?"

Another difficulty that the elephant might face along the way is that change, especially in terms of organisational culture issues like safety culture, is a long and challenging path. How can one continue to motivate the elephant? Employees must be convinced that there will be failures along the way, but these expected failures do not end the mission. This mentality is an important aspect of the growth mindset, which praises effort rather than natural skill and believes in the ability of people to grow and adopt new skills along the way. Thus, a learning culture (as highlighted in Chapter 7), which includes the belief that people can change and improve, is critical to building a safety culture. When incidents and accidents occur as a company embarks on a change journey to improve safety culture and safety performance, the morale of employees may drop, especially those directly involved in the incidents or implementation of change. This can derail the whole change effort. Thus, employees must be prepared to face failures and learn from them.

8.3.2.3 *Shape the path*

To shape the path, organisations must tweak the environment so that it is supportive of the desired behaviour, build habits through action triggers, checklists or other aids, and make use of the herd effect to spread the behaviour that the change process is hoping to build.

There is a natural tendency to blame individuals for their error, but as discussed in earlier chapters, many-a-times, the errors or unsafe behaviours were induced by the environment that they were in. For example, in the past, Automated Teller Machines (ATMs) were designed to give you the money first before returning the card and this caused many people to forget to keep their ATM cards. A simple change in sequence where the ATM card is returned first before the money is given to you resulted in a significant drop in the chances of forgetting the ATM card. Tweaking the environment overlaps with the field of human factors and ergonomics, which design the environment and equipment to suit human characteristics, alleviate our limitations and minimise common errors.

Another simple but effective approach to improve safety behaviour is to ensure that the work environment is neat and tidy. When housekeeping is poor, when objects are placed near the edges of openings or buildings, or when there are slipping hazards, it is a contravention of Reg 24 and 26 of the WSH (Construction) Regulations 2007 (as an example). When workers work in an environment where WSH contraventions are the norm, they can easily form unsafe attitudes and there is a higher tendency to violate other WSH rules and standard operating procedures. On the other hand, if workers work in an environment where housekeeping is properly and consistently practiced, where equipment and materials are kept in the correct locations, the perceived norm would be that WSH rules and procedures are important and adhered to. This will lead to a change in the behaviour of individuals placed in this environment.

Shaping the path is also about creating habits that are automatic. To trigger these habits or routines, change leaders can develop action triggers such as posters, signage and hand signals, which can be incorporated into trainings. Workers can be trained on abbreviations like STOP (Stop, Think, Observe and Plan) during training. Complementary posters can be placed in the workplace to trigger the routine taught in training, which includes taking a pause during work to think about the possible hazards and controls in the current task and location, and to observe the task and location so that a plan can be put in place to ensure that the work can be conducted safely. These action-triggers need to be easy to understand and the actions required need to be carefully scripted as discussed earlier.

Another commonly used trigger is a checklist. A crane operator is required to conduct a check according to a checklist prior to operations. The checklist is simple, but it helps to ensure that the check is thorough and it prevents over-confidence. However, to ensure that the checklist is used properly, the operator must understand its importance and supervisors must inspect the completed checklists every day and ask probing questions regarding the checklist. If the crane operators (i.e. the elephant in the operators) feel that the checklists are for novices, then the checklists will not be used even if they are useful. One way to minimise this is to show that true professionals use checklists; think of pilots!

The last approach to shape the path is to rally the herd or to make safe behaviour contagious. This idea overlaps with earlier ideas but is focused on the behavioural norms of the target group you are trying to change. Individuals tend to behave in accordance to the norm in a group. This has been supported by research such as Goh and Binte Sa'adon (2015), who found that the norm as perceived by workers has strong influence on the intention and actual behaviour of a worker hooking on his safety harness.

The elephant is always looking to the herd for cues about how to behave. Workers with unsafe behaviours can be identified through inspections and behavioural observations. Once these workers are identified, supervisors and change leaders can get them to change by assigning buddies with safe behavioural habits to the worker to remind the worker to work safely at the beginning of and during shifts. The intention is to work through the smallest possible group to influence the individual's behaviour, to bring about positive cultural change. Concurrently, individuals identified for their safe behaviour will need opportunities to share their experience in implementing the desired change and supporting each other in the effort.

Heath and Heath's (2010) Rider-Elephant-Path model is a compilation of different research on change management. The model is simple and practical for change at individual and different organisational levels. Not all the nine approaches highlighted are always useful, and overlaps exist between the different approaches, so change leaders need to consider the situation and use the model selectively.

8.3.3 Senge's change model

Another useful piece of literature on change management is Senge (1999), which highlighted three basic virtuous cycles that need to be

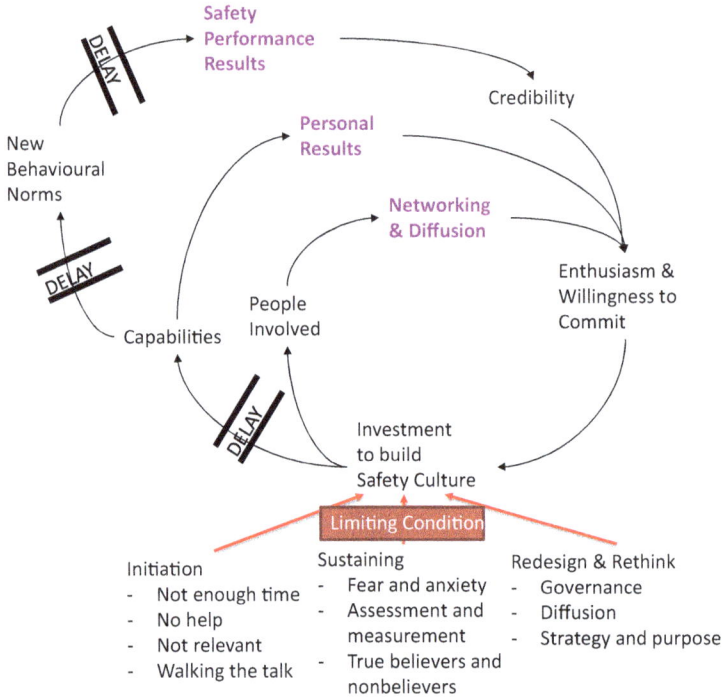

Figure 8.2   Virtuous cycles and limiting conditions; adapted from Senge *et al.* (1999)

in place for organisational change to be implemented and sustained (Figure 8.2):

1. Organisation's Safety Performance Results — the investment to build safety culture will take some time (delay) before capabilities in SHE management and risk controls can be improved. After a sustained period, the behaviours of people in the organisation will be improved and it will become a new norm for people to work safely. Since unsafe acts and unsafe conditions will be reduced, managers should expect to see an improvement in WSH performance. The improvement will improve the credibility of the initial investments and interventions. This will then encourage more investment in the programme and this will spark off a virtuous cycle of improvement.

2. Networks of Committed People — The initial investment to build safety culture will get people involved in improving WSH. The people that are directly involved will typically be more committed, but their initial numbers may be small. Getting these early supporters to network and diffuse their commitment and support will help to promote enthusiasm for the programme. This will encourage stronger commitment to the safety culture programmes and further investment of resources and time. This virtuous cycle will sustain the programme and keep people engaged.

3. Personal Results — The initial investment to build safety culture will increase WSH capabilities and after some time, the organisation should ensure that individuals see personal results from their improved WSH management capabilities. This means that the way that organisations measure and reward performance needs to encourage good safety behaviour. Once individuals see positive consequences from their involvement in the safety culture programme, they will be more enthusiastic and more willing to commit to the programme. This will then develop into a virtuous cycle at the personal level.

All three virtuous cycles will help organisations to reap the benefits of their safety culture improvement programme. Managers trying to get organisations to change their safety culture or implement organisational changes will need the help of these virtuous cycles.

However, for the organisational change to be successful, it is critical for the organisation to remove barriers to the three virtuous cycles (see Figure 8.2). The barriers and solutions identified by Senge *et al.* (1999) are briefly described in Table 8.1.

8.4 Change Management Plan

It is useful to have a change management plan with detailed documentation to guide and manage the organisational change process. An example

Table 8.1   Summary of the key change barriers and solutions; adapted from Senge *et al.* (1999)

Barriers or Challenges	Possible Solutions
1. "We don't have time for this stuff" • Lack of control over one's time. People involved in change initiatives need enough flexibility to devote time to reflection and practice • People might be committed, but their lack of time creates an inability to meet the demands of the change initiative • End up frustrated and giving up on the initiative • Especially true for pilot team members initiating change	**Initiation** • Integrate initiatives • Proper scheduling • Trust people to use their own time • Value unstructured time • Build productivity capability • No politics or games-playing and nonessential demands
2. "We have no help!" • Inadequate coaching, guidance and support for innovating group • No assistance to help build capacity to sustain change	• Invest in help — consultants, internal experts, guides, etc. • Creating capacity for coaching • Find a partner to confide in • Building coaching into line management • Treat seeking help as norm and positive
3. "This stuff isn't relevant!" • The absence of a clear, compelling business case for learning • Resulting in lack of commitment and enthusiasm	• Build awareness among key leaders • Explicitly raise questions about relevance in the pilot group • Make more information available to pilot group members — connect to individual and job goals • Keep training linked tightly to business results. • Revisit relevance periodically

4. "They're not walking the talk!" • Lack of clarity and credibility of the management's aims and values • Leads to lack of trust in management to carry out change	• Develop espoused aims and values that are credible • Demonstrate the values and commitment to aims through actions • Work with partners • Cultivate patience under pressure and with bosses • Develop a greater sense of organisational awareness (go to the ground) • Reflect on your beliefs about people and discuss values

Sustaining

5. Fear and anxiety — Common questions: • Am I safe? • Am I adequate? • Can I trust myself and others to sustain the change?	• Start small and build momentum before confronting difficult issues • Avoid frontal assaults • Set an example of openness • Learn to see diversity as an asset • Use defects as opportunities for learning • Make sure participation in pilot groups and change initiatives is by choice, not by coercion • Remember fear and anxiety are natural; forcing creates more anxiety
6. Assessment and measurement • Expectations of results • Delay in seeing results • Negative results due to change • Lack of measurement of results	• Appreciate time delays in change initiatives • Assign suitable metrics (not just cost or final outcome) • Recognise and appreciate progress

(Continued)

Table 8.1 (*Continued*)

Barriers or Challenges	Possible Solutions
Redesign and Rethink	
7. True believers vs. nonbelievers (lack of understanding between groups) • Pilot group (believers) vs. those not in the pilot (nonbelievers) • Fear and anxiety arises among those not involved in the change • Lack of engagement between groups • Arrogance among pilot group members who think they are always right • Pilot group might feel under-appreciated for their efforts	• Become "bicultural" — incubators and mainstream • Mentoring of line leaders • Build ability to engage so as to spread change initiative beyond pilot group • Cultivate reflective openness • Respect people's inhibitions about personal change • You don't have to convince people • Remove jargons and keep language simple • Lay foundation of transcendent values (tolerance and flexibility)
8. Governance ("Who is in charge of this stuff?") • How to institutionalise the change initiatives to be aligned with existing structures • Balance between autonomy for change and corporate control	• Pay attention to boundaries and be strategic when crossing them • Articulate the case for change in terms of business results • Make executive leaders' priorities part of your creative thinking • Experiment with cross-functional, cross-boundary teams, if you can get them sponsored by the hierarchy • Begin at the beginning: with governing ideas • Deploy new rules and regulations judiciously • Never underestimate the power of small changes in complex situations — if they are the "right" changes • Be prepared for a long journey and don't embark alone

9. Diffusion ("Reinventing the wheel") • Not learning from pilot groups' experience • Other people's innovation were dismissed as not applicable to them • Lack of interest • Impatience in repeating change initiatives • Arrogance — Assume that there is nothing new to learn	• Increase quality and number of coaches • Increase permeability of intra organisational boundaries • Improve information infrastructure — share innovations and changes • Build learning culture & communities of practice • Research as part of executive accountability
10. Strategy and Purpose • Change can create questions about existing strategy and purpose • What do we really want to create? • How will it contribute to others? • New ideas on strategy and purpose may not be accepted	• Use scenario thinking to investigate blind spots, signals of unexpected events and organisational purpose • Engage people in organisational strategy and purpose • Expose and test assumptions behind strategy

can be found in the Queensland Government Chief Information Office (n.d.), which includes the following components:

1. Change identification: Type, reason, scope, current status, desired state
2. Change specification: policy, structure, processes, relationships, culture, people, information, cost, risk assessment
3. Change methodology: Stakeholder analysis (target groups, advocates, early adopters, resistance and drivers)
4. Implementation strategies: Action plan, schedules, communication plan, training plan, IT/business systems plan, resistance management plan

8.5 Case Study: Alcoa

This case study is adapted from Duhigg (2014) (Chapter 4), which describes how the Aluminium Company of America, Alcoa, successfully established a culture of excellence by focusing on safety as a keystone habit. Alcoa is a global industrial leader that has businesses in different industries. In 1987, Paul O'Neill, was appointed as the new CEO. He was a former government bureaucrat and was relatively unknown in the Wall Street community. O'Neill made a shocking announcement in his first shareholder meeting. He indicated that he intended to make Alcoa the safest company in America and he wanted to go for zero injuries. This was unprecedented and investors in the room were shocked.

O'Neill believes that bringing down the injury rates would mean that individuals in the company have agreed to devote themselves to creating a habit of excellence. In this way, safety becomes an indicator of how good the company is in changing shared habits. O'Neill wanted to attack one habit and then let the effect of the change ripple through the organisation. This is what Duhigg (2014) defined as a "keystone habit", a habit that has the power to start a chain reaction to influence how all employees behave and communicate in the organisation. These keystone habits are a focus and leverage for managers to initiate fundamental changes in organisational culture. Instead of framing safety culture as a subset of organisational culture, safety culture becomes a trigger for change in organisational culture.

O'Neill required managers to understand why injuries happen. In the process of investigating incidents, managers had to understand how the manufacturing process went wrong and to understand and rectify the process problems. Workers needed to be educated about quality control, efficient work processes, how to make processes less error-prone, etc. The entire campaign centred around the idea that correct work is safe work and vice versa.

O'Neill's commitment was put to the test when a fatality occurred. An extrusion press (a metal working machine) had stopped operating; a worker jumped over the yellow safety wall surrounding the press and walked across the pit to remove a piece of aluminium jammed into the hinge of a swinging six-foot arm. When the jammed aluminium scrap was removed by the young man, the machine jumped back into motion, and the mechanical arm swung and struck the worker on his head, crushing his skull and killing him. The worker was a young male employee who had just joined the company a few weeks ago. The worker was eager to take up the job because his wife was pregnant and they needed the health care the job offered.

O'Neill was informed about the accident in the middle of the night and within 14 hours, ordered all the plant's executives and Alcoa's top officers to its headquarters to attend an emergency meeting. They reviewed the evidence and analysed the accident in detail. They identified several causes including:

- two managers saw the worker jump over the barrier but failed to stop him
- a training programme that failed to emphasise to the man that he would not be blamed for a breakdown

- lack of instructions that he should find a manager before attempting a repair
- absence of sensors to automatically shut down the machine when someone stepped into the pit

The CEO said, "We killed this man… It's my failure of leadership… And it's the failure of all of you in the chain of command." This is a very strong statement, but more indicative of his commitment were his subsequent actions. O'Neill gave his home telephone number to workers on the ground, asking them to give him a call whenever there are safety issues that their managers did not follow up on. According to Duhigg (2014), workers started to call and they were giving him great ideas to improve work in general.

One example is when a worker made a suggestion to improve the work process at an aluminium manufacturing plant that manufactured siding for houses. The plant had engaged consultants to choose shades of paint to anticipate the colours that customers would want, but despite the hefty consultant fees, the problem remained unsolved. The worker suggested that all the painting machines be grouped together, so that paint pigments could be switched out faster, allowing them to be nimbler in responding to changes in customers' colour preferences. This simple suggestion resulted in profits on aluminium siding within a year. Interestingly, the worker had had the idea for a decade, but he had not told his management. Instead, it was only when O'Neill started focusing on improving safety, that the communication channel was opened — and that was when the employee thought that his suggestion might be useful, and finally decided to raise it. The focus on improving safety culture created a climate for innovation and willingness to change.

By 2000, when O'Neill retired, the annual net income had increased by five times as compared to when O'Neill initially started. Market capitalisation had risen by US$27 billion and Alcoa had become one of the safest companies in the world. More specifically, prior to 1987, the Alcoa plant had at least one accident per week, but after O'Neill took over as CEO, there were plants that had not had accidents that caused more than one day of work to be lost. The company's worker injury rate was one-twentieth of the US average.

This case study shows that improving safety culture is possible, and Alcoa is a bright spot that should be cloned. The importance of change leaders cannot be overstated, and it is obvious that the change was difficult, but the benefits go beyond WSH.

8.6 Conclusions

Improving safety culture is not an easy task. Managers need to structure the change systematically and deliberately. This chapter highlighted several approaches for improving safety culture. Safety culture intervention is a specific type of organisational change; hence it is useful to understand a range of organisational change management models. The change management models introduced include Lewin's model, Heath and Heath's (2010) Rider-Elephant-Path model, and Senge *et al.*'s (1999) virtuous cycles and barriers. A template for a change management plan was also briefly discussed. Lastly, Alcoa's successful change initiative was presented as a bright spot of a positive safety culture and how organisational excellence can be improved.

Review Questions

1. What are the three general approaches to improving safety culture?
2. Describe Heath and Heath's (2010) Switch model of change.
3. Describe the three virtuous cycles needed in making sustainable changes in organisations.
4. According to Senge *et al.* (1999), what are the 10 types of barriers to change?
5. Apply the Rider-Elephant-Path model to Alcoa's case study. What are the approaches used to implement the change?

References

Duhigg, C. (2014). *The power of habit: why we do what we do in life and business*, Random House, New York.

Goh, Y. M., and Binte Sa'adon, N. F. (2015). "Cognitive factors influencing safety behavior at height: A multimethod exploratory study." *Journal of Construction Engineering and Management*, 141(6).

Heath, C., and Heath, D. (2010). *Switch: how to change things when change is hard*, Broadway Books, New York.

Queensland Government Chief Information Office (n.d.). "Change Management Plan Workbook and Template." <http://www.nrm.wa.gov.au/media/10528/change_management_plan_workbook_and_template.pdf>. (March 12, 2018).

Senge, P. (1999). *The dance of change: the challenges of sustaining momentum in learning organizations: a fifth discipline resource*, Nicholas Brealey Pub, London.

Chapter 9 Overview of WSH Legislations

9.1 Introduction

In Singapore, workplace safety and health (WSH) management are regulated by a suite of regulations. As part of any WSH management system, the organisation needs to have procedures to identify and access relevant WSH legislations and ensure compliance to the regulations. This chapter provides a brief overview of the Workplace Safety and Health Act (WSHA) and its subsidiary legislations. This chapter relies heavily on the content provided on the Ministry of Manpower (MOM) website and Singapore Statutes Online.

Table 9.1 contains the list of legislations under the Workplace Safety and Health Act (Cap. 354A) of Singapore. The WSHA was discussed in Chapter 1. In addition, some of the subsidiary regulations, such as the WSH Design for Safety Regulations 2015 (S428/2015) and the WSH (Risk Management) Regulations (Cap. 354A, RG 8), were also discussed in the relevant chapters. Other details of the WSHA and selected regulations will be briefly discussed in this chapter.

9.2 Workplace Safety and Health Act (1ˢᵗ Jan 2018 Version)

The key duties of the different duty holders were discussed in Chapter 1. The following will provide additional information highlighted in the Workplace Safety and Health Act (WSHA).

Table 9.1   List of WSHA-related legislations in Singapore

#	Title	Number
1	Workplace Safety and Health Act	Cap. 354A
2	Factories (Registration and Other Services — Fees and Forms) Regulations	Cap. 354A, RG 5
3	Factories (Safety Training Courses) Order	Cap. 354A, OR 12
4	Factories (Work of Engineering Construction) Order	Cap. 354A, OR 6
5	Workplace Safety and Health (Abrasive Blasting) Regulations 2008	S 607/2008
6	Workplace Safety and Health (Asbestos) Regulations 2014	S 337/2014
7	Workplace Safety and Health (Composition of Offences) Regulations	Cap. 354A, RG 6
8	Workplace Safety and Health (Confined Spaces) Regulations 2009	S 462/2009
9	Workplace Safety and Health (Construction) Regulations 2007	S 663/2007
10	Workplace Safety and Health (Design for Safety) Regulations 2015	S 428/2015
11	Workplace Safety and Health (Exemption) Order	Cap. 354A, OR 1
12	Workplace Safety and Health (Explosive Powered Tools) Regulations 2009	S 325/2009
13	Workplace Safety and Health (First-Aid) Regulations	Cap. 354A, RG 4
14	Workplace Safety and Health (General Provisions) Regulations	Cap. 354A, RG 1
15	Workplace Safety and Health (Incident Reporting) Regulations	Cap. 354A, RG 3
16	Workplace Safety and Health (Major Hazard Installations) Regulations 2017	S 202/2017
17	Workplace Safety and Health (Medical Examinations) Regulations 2011	S 516/2011
18	Workplace Safety and Health (Noise) Regulations 2011	S 424/2011
19	Workplace Safety and Health (Offences and Penalties) (Subsidiary Legislation under Section 66(14)) Regulations	Cap. 354A, RG 5
20	Workplace Safety and Health (Operation of Cranes) Regulations 2011	S 515/2011
21	Workplace Safety and Health (Registration of Factories) Regulations 2008	S 501/2008

(Continued)

Table 9.1   *(Continued)*

#	Title	Number
22	Workplace Safety and Health (Risk Management) Regulations	Cap. 354A, RG 8
23	Workplace Safety and Health (Safety and Health Management System and Auditing) Regulations 2009	S 607/2009
24	Workplace Safety and Health (Scaffolds) Regulations 2011	S 518/2011
25	Workplace Safety and Health (Shipbuilding and Ship-Repairing) Regulations 2008	S 270/2008
26	Workplace Safety and Health (Transitional Provision) Regulations	Cap. 354A, RG 7
27	Workplace Safety and Health (Work at Heights) Regulations 2013	S 223/2013
28	Workplace Safety and Health (Workplace Safety and Health Committees) Regulations 2008	S 355/2008
29	Workplace Safety and Health (Workplace Safety and Health Officers) Regulations	Cap. 354A, RG 9
30	Workplace Safety and Health (Workplaces Subject to Act) Order 2007	S 72/2007

9.2.1 Difference between workplace and factory

The WSHA provided a distinction between a "workplace" and a "factory". A "workplace" means any premises where a person is at work or is to work, for the time being works, or customarily works, and includes factories. Some of these workplaces are further classified as a factory. A factory is a term carried over from the repealed Factories Act, which was the main legislation regulating industrial safety and health. A factory is any premises in which any of the following is carried out:

(a) The making of any article or part of any article.
(b) The alteration, repair, ornamentation, finishing, cleaning or washing of any article.
(c) Breaking up or demolishing any article.
(d) Adapting any article for sale.

As described in the Fourth Schedule of the WSHA, examples of factories include manufacturing plants, car-servicing workshops, shipyards and construction work-sites. These workplaces are considered higher risk and are subjected to additional regulations, e.g. periodic audits and a mandatory WSH committee.

9.2.2 Protecting employees

The WSHA Section 18 protects employees by stating that employers shall not "(1)(a) deduct, or allow to be deducted, from the sum contracted to be paid by him to any employee of his; or (b) receive, or allow any agent of his to receive, any payment from any employee of his, in respect of anything to be done or provided by him in accordance with this Act in order to ensure the safety, health or welfare of any of his employees at work." Anecdotally, there are practices in the industry where employers (including principals) deduct the pay of employees or ask for payment for personal protective equipment that should have been provided to the workers as required by WSHA. In addition, Section 18 also make it illegal for employers to dismiss or threaten to dismiss an employee if the employee whistle-blows on the employer for WSHA-related contraventions, or for performing his duties as a member of the WSH committee.

9.2.3 Powers of commissioner

The WSHA gives the Commissioner for WSH, who is typically the Director of the Division of Occupational Safety and Health in the Ministry of Manpower, powers to issue stop work orders (to stop the work at a workplace) and remedial orders (to improve the conditions in a workplace). Failure to comply with a remedial order can result in a fine not exceeding S$50,000 or imprisonment for a term not exceeding 12 months or both. In the case of a continuing offence, the duty holder may have to pay a further fine not exceeding S$5,000 for every day or part thereof during which the offence continues after conviction. On the other hand, failure to comply with a stop work order can result in a fine not exceeding S$500,000 or imprisonment for a term not exceeding 12 months or both. In the case of a continuing offence, the duty holder may have to pay a further fine not exceeding S$20,000 for every day or part thereof during which the offence continues after conviction.

The Commissioner also has the power to enter a workplace for the purpose of inspection and investigation. The WSHA also made provisions to protect accident-related evidence, where any alteration or addition to machinery, equipment, etc. related to an accident and dangerous occurrence is an offence.

Section 27A of the WSHA is a proactive amendment of the WSHA in 2018. The section allows the Commissioner to prepare and publish a learning report on any accident, dangerous occurrence or occupational disease in a workplace that is still being investigated. Prior to this section, it was difficult for MOM to disseminate important WSH information gained from investigations prior to prosecution because the information released may influence the outcome of the prosecution. The learning report may:

(a)   contain an account of the accident, dangerous occurrence or occupational disease;
(b)   specify the cause or causes of, and circumstances or factors leading to, the accident, dangerous occurrence or occupational disease insofar as they may be ascertained;
(c)   contain an opinion by a person with technical or specialised knowledge of the machinery, equipment, plant, article, process, substance, work or workplace involved in the accident, dangerous occurrence or occupational disease;
(d)   contain a warning of any danger or risk to the safety and health of persons at work or persons who may be affected by any undertaking carried on in the workplace;
(e)   contain any recommendation to prevent or minimise the recurrence of any similar accident, dangerous occurrence or occupational disease in a workplace; and

(f)   contain any other matter that the Commissioner considers relevant, taking into account the sole objective mentioned in subsection (2).

The learning report is not admissible in any legal proceedings. This is aligned with the approach used by the US Chemical Safety Board (CSB), where the investigation report written by CSB is not admissible in court and the US Occupational Safety and Health Administration (OSHA) conducts an independent investigation for its prosecution purposes.

9.2.4 Safety and health management arrangements

The WSHA Part VII specifies requirements for safety and health management arrangements. These requirements pertain to WSH Officers (WSHOs) and co-ordinators, WSH committees, WSH auditors, safety and health training courses and the need for the Commissioner's approval for people to perform statutory roles (authorised examiner for dangerous machines, WSHO, WSH co-ordinator, WSH auditor and accredited training provider). The WSHA only provided the framework to regulate each of these areas of safety and health management arrangements. The details of the actual requirements are usually found in relevant subsidiary regulations and government gazettes.

9.2.5 WSH council and approved code of practice

The WSHA also provided for the establishment of the WSH Council. According to Section 40A, the functions of the WSH Council are:

(a)   to develop or facilitate the development of acceptable practices relating to safety, health and welfare at work;

(b)   to promote the adoption of acceptable practices relating to safety, health and welfare at work;

(c)   to devise, organise and implement programmes and other activities for or related to providing support, assistance or advice to any person or organisation in preserving, improving and promoting safety, health and welfare at work;

(d) to facilitate and promote the development and upgrading of competencies, skills and expertise of the workforce relating to safety, health and welfare at work;

(e) to research into any matter relating to safety, health and welfare at work;

(f) to grant prizes and scholarships, and to establish and subsidise lectureships in universities and other educational institutions in subjects relating to safety, health and welfare at work;

(g) to provide practical guidance with respect to the requirements of this Act relating to safety, health and welfare at work; and

(h) to do all the things that it is authorised or required to do under the WSH Act.

Sections 40B and 40C then stipulated the role of the codes of practice as issued or approved by the WSH Council. The codes of practice approved by the WSH Council are frequently Singapore Standards published by the Standards, Productivity and Innovation Board under Section 7(2)(h) of the Standards, Productivity and Innovation Board Act (Cap. 303A). Essentially, the approved code of practice (ACoP), is a benchmark for the court to determine whether a measure is reasonable and practicable. In the absence of a suitable ACoP, relevant Singapore Standards, not approved by the WSH Council and industry guidelines, can also be used to establish standards that are reasonable and practicable.

9.2.6 Schedules

The First Schedule provided the specific types of dangerous occurrences which must be reported to the Ministry of Manpower. They include bursting failures of machinery, crane collapses, significant explosions or fires, significant electrical short-circuits or failures, explosions or failures of a steam boiler's structure(s), failures or collapses of formwork or its supports, scaffold (15 metres in height) collapses of suspended or hanging scaffold above two metres, and accidental seepage or entry of seawater into a dry dock.

The Second Schedule identified 35 occupational diseases that must be reported to the Manpower Ministry:

1. Aniline poisoning
2. Anthrax
3. Arsenical poisoning
4. Asbestosis
5. Barotrauma
6. Beryllium poisoning
7. Byssinosis
8. Cadmium poisoning
9. Carbamate poisoning
10. Compressed air illness or its sequelae, including dysbaric osteonecrosis
11. Cyanide poisoning
12. Diseases caused by ionising radiation
13. Diseases caused by excessive heat
14. Hydrogen Sulphide poisoning
15. Lead poisoning
16. Leptospirosis
17. Liver angiosarcoma
18. Manganese poisoning
19. Mercurial poisoning
20. Mesothelioma
21. Noise-induced deafness
22. Occupational asthma
23. Occupational skin cancers
24. Occupational skin diseases
25. Organophosphate poisoning
26. Phosphorus poisoning
27. Poisoning by benzene or a homologue of benzene
28. Poisoning by carbon monoxide gas
29. Poisoning by carbon disulphide
30. Poisoning by oxides of nitrogen
31. Poisoning from halogen derivatives of hydrocarbon compounds
32. Musculoskeletal disorders of the upper limb
33. Silicosis
34. Toxic anaemia
35. Toxic hepatitis.

The Third Schedule provided a list of activities classified as Work of Engineering Construction, i.e. construction work, while the Fourth Schedule identified 19 categories of workplaces defined as factories. The Fifth Schedule listed 11 types of machinery and equipment (e.g. scaffolds, cranes, forklifts, bar benders and welding machines) and 17 types of hazardous substances (e.g. corrosive, flammable, explosive, oxidising, toxic and carcinogenic substances) that require their manufacturers and suppliers to ensure compliance with Section 16 of WSHA, which makes it compulsory to:

(a) Provide information on health hazards and how to safely use the machinery, equipment or hazardous substance.
(b) Examine and test the machinery, equipment or hazardous substance to ensure that it is safe for use.
(c) Provide results of any examinations or tests of the machinery, equipment or hazardous substances.

The persons who erect, install or modify the machinery and equipment identified in the Fifth Schedule are required under Section 17 to ensure that the method of work is in accordance to the manufacturer's instruction and the machinery or equipment is safe for use after installation, erection or modification.

9.3 Workplace Safety and Health (Asbestos) Regulations 2014

Asbestos is a hazardous substance that can cause lung diseases such as asbestosis (scarring and fibrosis of the lung tissues), mesothelioma (a cancer of the chest and abdominal lining) and lung cancer (WSH Council, 2017). The substance is not a specific mineral, but a group of fibrous silicate minerals (Tranter, 2004) that was commonly used for insulation, fire protection and acoustic purposes. The WSH (Asbestos) Regulations 2014 defines asbestos as crocidolite, actinolite, anthophyllite, amosite, tremolite, chrysotile, amphiboles or a mixture containing any such minerals. The material has been banned in Singapore since 1989, but it can still be found in the thermal insulation of pipes in vessels, plants and furnaces, and buildings constructed before 1989. The National Environment Agency (NEA) and the Ministry of Manpower regulates the use and disposal of asbestos in Singapore. MOM is focused on the protection of people at work and the key legal instrument is the WSH (Asbestos) Regulations 2014 ("Asbestos Reg").

The Asbestos Reg contains six parts and a Schedule. The Schedule defines the material, substance, product or article classified as "specified material" that requires special attention and measures under the Asbestos Reg. The Schedule listed the following "specified material":

1. Cable penetration insulation
2. Fire protection board, panel, wall and door
3. Gasket
4. Refractory lining
5. Sprayed insulation
6. Thermal insulation of pipe, boiler, pressure vessel and process vessel.

For any work that involves the above specified material that is likely to contain asbestos, or any demolition, alteration, addition or repair of a building built before 1 Jan 1991, the employer or principal must engage a competent person to assess if the specified material contains asbestos. The competent person will then conduct a survey to collect suitable samples and send it for testing by an approved testing body. A survey report must then be provided after the test with regards to the presence of asbestos.

If asbestos is present in the specified material, then the Occupier must ensure that the asbestos removal work is conducted by an Approved Asbestos Removal Contractor (AARC) (a list of AARCs can be found on the MOM website). Work involving asbestos must be conducted by workers who are trained to understand the hazards of asbestos and the safe way to handle the material, including training on the use of respiratory protective equipment. These workers will also be subjected to medical examinations stipulated in the Workplace Safety and Health (Medical Examinations) Regulations 2011. The AARC must engage a competent person to develop an asbestos-removal plan of work and supervise the asbestos-removal work. The AARC must also ensure implementation of the plan. MOM must be notified of all asbestos removal work seven days prior to the commencement of the removal work.

9.4 Workplace Safety and Health (Confined Spaces) Regulations 2009

According to the WSH (Confined Spaces) Regulations 2009 ("Confined Spaces Reg"), a confined space is defined as any chamber, tank, manhole, vat, silo, pit, pipe, flue or other enclosed space, in which:

(a)   dangerous gases, vapours or fumes are liable to be present to such an extent as to involve a risk of fire or explosion, or persons being overcome thereby;

(b)   the supply of air is inadequate, or is likely to be reduced to be inadequate, for sustaining life; or

(c)   there is a risk of engulfment by material.

Confined spaces are dangerous because they may not be easily identified by workers. There have been accidents where multiple workers were killed because when the first worker become unconscious in the confined space, co-workers entering the confined space to rescue the affected worker also become affected by the harmful gases or substances in the space. The Confined Spaces Reg stipulates the requirement for the Occupier to record the description and location of all fixed and stationary confined spaces and inform the relevant people of the hazards of these identified confined spaces. The Occupier must also ensure that the access and egress, entrance, lighting and ventilation of the confined spaces are safe and suitable for the confined spaces.

The Confined Spaces Reg stipulates a permit-to-work system where a competent authorised manager and a competent confined space safety assessor are appointed. The confined space entry permit must specify the following:

(a) the description and location of the confined space;
(b) the purpose of entry into the confined space;
(c) the results of the gas testing of the atmosphere of the confined space; and
(d) its period of validity.

The Occupier is to evaluate if the confined space entry is indeed necessary. If the entry is necessary, the Occupier must ensure that the permit-to-work system is implemented. If the person entering the confined space is wearing a suitable breathing apparatus, authorised by the authorised manager to enter the confined space, and "where reasonably practicable, is wearing a safety harness with a rope securely attached and there is a confined space attendant keeping watch outside the confined space who is provided with the means to pull such person out of the confined space in an emergency", then the confined space entry permit is not necessary.

The application for the confined space entry permit is to be made by the supervisor of the person entering the confined space. The submitted application form will have to contain the safety measures to protect the workers entering the confined space. These safety measures will have to be assessed by the confined space assessor and then approved by the authorised manager before the workers can enter the confined space. The entry permit must be displayed at the entrance of the confined space.

One of the duties of the confined space entry assessor is to test the atmosphere of the confined space for oxygen content (19.5% to 23.5% by volume), level of flammable gases or vapour (less than 10% of its lower explosive limit), and the presence of toxic gas or vapour (that it does not exceed the permissible exposure levels as specified in the First Schedule to the Workplace Safety and Health (General Provisions) Regulations

(Rg 1)). The gas meter used in the testing must be properly calibrated. The confined space safety assessor will have to conduct periodic testing of the atmosphere during the conduct of the work. In addition, if there is more than one person present in the confined space, at least one of them must have a suitable gas detector that continuously monitors the atmosphere in the confined space. In the event that a hazardous atmosphere is detected, all personnel must vacate the confined space. A re-evaluation of the confined space must be conducted, and a new permit will have to be issued before any worker can enter the confined space.

Another key hazard of a confined space is "incompatible work", which means work which is carried out at or in the vicinity of any work carried out in the confined space and which is likely to pose a risk to the safety and health of persons present in the confined space. A set of incompatible work includes hot work (e.g. welding, hot-cutting and grinding) and spray painting in confined spaces. The work may be in adjacent confined spaces, but when the incompatible substances come into contact a major fire and explosion can occur. Thus, the Confined Spaces Reg requires anyone who is aware of incompatible work to immediately report it to their supervisor, WSH personnel and authorised manager.

9.5 Workplace Safety and Health (Construction) Regulations 2007

The WSH (Construction) Regulations 2007 ("Con Reg") is one of the core WSH regulations for the construction industry. It contains 16 parts and 142 Regulations.

In addition to the WSH (Safety and Health Management System and Auditing) Regulations 2009, the Con Reg requires the Occupier to convene site coordination meetings to coordinate site activities so as to ensure the safety, health and welfare of persons at work in the worksite. The site coordination meeting has to be chaired by the "project manager", which means the person who is stationed at a worksite and who has overall control of all the works carried out in the worksite, and includes any competent person appointed by the occupier of the worksite in the event that the project manager is unable to perform his duties under the Con Reg. In addition, the Occupier of a project with contract sum less than $10 million will have to appoint a WSH co-ordinator who is meant to assist the occupier in identifying unsafe conditions or unsafe work practices, recommend to the occupier reasonably practicable measures to remedy the unsafe conditions or unsafe work practices, and assist the occupier in implementing the recommended reasonably practicable measures. This role is similar to that of a WSH Officer, but the level of training of a WSH co-ordinator is of a lower level than a WSH Officer. The Con Reg also requires suitable safety and health training for all workers and supervisors.

Similar to the Confined Spaces Reg, the Con Reg also requires a permit-to-work system for the following high-risk construction activities:

(a)   demolition work;
(b)   excavation and trenching work in a tunnel or hole in the ground exceeding a depth of 1.5 metres;
(c)   lifting operations involving a tower, mobile or crawler crane;
(d)   piling work; and
(e)   tunnelling work.

The remaining portions of the Con Reg cover different hazards, such as structural safety and stability, storage and stacking of materials and equipment, falling objects, slipping hazards, protruding hazards, personal protective equipment, electrical safety, cantilevered platforms, formwork, demolition, excavation and tunnelling, piling, crane and employee's lifts.

9.6 Workplace Safety and Health (General Provisions) Regulations

The WSH (General Provisions) Regulations ("GP Reg") covers all workplaces. It covers general provisions relating to health, safety and welfare. The hazards identified under health issues include: infectious agents and biohazardous material, overcrowding, ventilation, lighting, drainage (to prevent wet floors), sanitary conveniences, vibration and excessive heat or cold and harmful radiations.

The range of safety hazards covered in the GP Reg is much wider. Several of the Regulations cover machinery hazards. For example, Regulations 11 stipulates the requirements for occupier to ensure secure fencing (or guarding) for prime movers (e.g. electric motor and internal combustion engines) and connecting flywheels and moving parts. Regulations 12 further indicates that fencing is not necessary only if the dangerous parts of the machinery are made safe (i.e. no person can be harmed) because of its position, construction or other means. Another requirement is the need for an emergency cut off that can efficiently cut off the power to the machinery. Regulations 13 requires safety measures to be implemented to assure the safety of maintenance workers who may need to remove the machinery fencing while the machinery is operating. The worker must be at least 18 years old. He must have been trained for maintenance or repair jobs and he must be wearing suitable clothing that does not have loose ends.

Other safety requirements include electrical installation and equipment (prevention of electrical shock), and requirements to ensure fencing and other safeguards are properly constructed, used and maintained. The occupier also needs to set up a set of lock-out procedures to prevent accidental activation or energising during the inspection, cleaning, repair or maintenance of any plant, machinery, equipment or electrical installation in the workplace. The Regulations also cover the danger of a person falling into the tank, structure, sump or pit containing scalding, burning, corrosive or toxic liquid. Self-acting machines (e.g. robotic arms) are required to be made safe by, for example, ensuring no person can be exposed to the machine in operation, and having warning signs.

Regulations 19 to 21 (hoists and lifts, lifting gears, lifting appliances and lifting machines) are related to the role of an Authorised Examiner (AE), who is essentially a special class of Professional Engineers authorised by the MOM to conduct the inspection and certification of dangerous lifting-related machineries. The occupiers and AEs play important roles in preventing accidents like lifted objects falling from height and failure of hoists.

Even though there is a WSH (Work At Heights) Regulations 2013 ("WAH Regs"), the GP Regs contains several requirements on fall from height hazards. Regulations 23 imposes a duty on the occupier to ensure that all openings in the floors of the workplace are securely covered or fenced unless the fencing is impracticable. Regulations 23 requires employers to provide "secure foothold(s) and handhold(s)", as far as is reasonably practicable, for any worker working at 2 metres or higher or if the worker can fall into any substance that can cause drowning or asphyxiation. If secure footholds and handholds are not reasonably practicable then the employer must provide other suitable means such as safety harnesses or safety belts (i.e. fall arrest or travel restraint systems). It was stipulated that the anchorage for the harnesses or safety belts must not be lower than the level of the working position of the worker.

Other safety hazards covered included storage of goods, explosives, flammable dust, gas, vapours or substances, pressure vessels (e.g. steam boiler, steam receiver, air receiver and refrigerating plant, which require AE inspection and certification) and fire.

Part IV covers toxic dust, fumes or other contaminants, permissible exposure levels of toxic substances, hazardous substances, warning labels and safety data sheets. It is noted that the occupier must ensure that any hazardous substance found in a workplace must have a safety data sheet. The First Schedule of the GP Regs shows the permissible exposure levels (PELs) for a wide range of hazardous hazards.

9.7 Workplace Safety and Health (Incident Reporting) Regulations

Under the WSH (Incident Reporting) Regulations ("IR Regs") the employer is responsible for notifying the Commissioner of Workplace Safety and Health (in practice this will be through the Ministry of Manpower) of any fatal accident involving its employees as soon as is reasonably practicable. The occupier will have to report the death of non-workers and the self-employed to the Commissioner. The report must be made within 10 days of the accident. If a dangerous occurrence occurs (see First Schedule of the WSH Act), the occupier shall, as soon as is reasonably practicable, notify the Commissioner of the occurrence. This must be done not later than 10 days after the occurrence.

In the event of an injury, the employer must submit a report to the Commissioner if the employee is granted more than three days of sick leave (consecutive or otherwise) not later than 10 days after the third day of the sick leave. Similarly, the employer must file a report, within 10 days, with the Commissioner if the employee is admitted to a hospital for at least 24 hours for observation or treatment. If the injury was sustained by a person who was not working or is self-employed, the occupier will have to notify the Commissioner.

In terms of occupational diseases listed in the Second Schedule in the WSH Act, the employer and registered medical practitioner shall each submit a report to the Commissioner within 10 days of the diagnosis.

The notification reports are legal documents and employers and occupiers must maintain a copy of the reports for at least three years.

9.8 Workplace Safety and Health (Operation of Cranes) Regulations 2011

Cranes are dangerous machines that can possibly kill many people during an accident. Thus, the Workplace Safety and Health (Operation of cranes) Regulations 2011 ("OOC Regs") identified a list of duties and requirements for the responsible person, which is frequently the occupier. One of the key administrative controls is the lifting plan. The lifting plan is described in more detail in the Code of Practice on Safe Lifting Operations in the Workplaces (Workplace Safety and Health Council, 2014). Cranes exceeding 5 tonnes and tower cranes can only be operated by a registered crane operator. Cranes not exceeding 5 tonnes (e.g. a mini crane or lorry crane) can only be operated by a competent operator, but the operator need not be registered with the Commissioner for WSH.

A registered crane operator is required to:

(a) conduct a pre-start inspection on the crane;
(b) ascertain the ground conditions and report to the lifting supervisor if he feels that it is not safe;
(c) ensure that any outrigger is fully extended and secured when it is required to;
(d) carry out lifting operations only if he has been briefed by the lifting supervisor on the lifting plan;
(e) ascertain the weight of the load, not lift when a signalman is required and a clear signal has not been given;
(f) operate the crane in a safe manner (e.g. no load over public area, no operation within 3 metres of a live overhead power line, no dragging or pulling of a load, and securely block a crane parked on a slope) and report any failure or malfunction of the crane to the lifting supervisor and make a record of the failure or malfunction in the crane's log book or log sheet.

The OOC Regs also identified other lifting personnel including the lifting supervisor, rigger and signalman. The lifting supervisor is to: coordinate all lifting activities, ensure only suitable and appointed personnel are involved in the lifting operations, ensure that the ground conditions are safe, brief all lifting personnel on the lifting plan and take measures to rectify all unsatisfactory or unsafe conditions to ensure that the lifting operation can be conducted safely.

Only approved crane contractors (ACC) are allowed to install, repair, alter or dismantle mobile cranes or tower cranes. ACCs need to be competent in their duties and only companies approved by the Commissioner for WSH can become an ACC. The ACCs need to be familiar with the crane manufacturers' instructions and manuals and they must follow the instructions and manuals strictly. The owner of the cranes must ensure that the crane is tested and certified by an AE.

9.9 Workplace Safety and Health (Work at Heights) Regulations 2013

Falling from height is the key reason for workplace fatalities. Thus, the WSH (Work At Heights) Regulations 2013 (WAH Regs) is an important legislation that all workplaces must consider. The WAH Regs follows the hierarchy of control, which advocates the importance of avoidance of WAH whenever possible. The occupier of the following workplaces must have a fall prevention plan:

(a)   Any worksite.
(b)   Any shipyard.
(c)   Any factory engaged in the processing or manufacturing of petroleum, petroleum products, petrochemicals or petrochemical products.
(d)   Any premises where the bulk storage of toxic or flammable liquids is carried on by way of trade or for the purpose of gain and which has a storage capacity of 5,000 or more cubic metres for such toxic or flammable liquids.
(e)   Any factory engaged in the manufacturing of —

    (a) fluorine, chlorine, hydrogen fluoride or carbon monoxide; or
    (b) synthetic polymers.

(f)   Any factory engaged in the manufacturing of pharmaceutical products or their intermediates.
(g)   Any factory engaged in the manufacturing of semiconductor wafers.
(h)   Any factory not falling within any of the classes of workplaces described in paragraphs 1 to 7, and in which 50 or more persons are employed.

The fall prevention plan is described in the Code of Practice for Working Safety at Heights (Workplace Safety and Health Council, 2013).

The responsible person (employer and/or principal) of any person conducting WAH must ensure that the person is trained to WAH. At the same time, the WAH must be under the "immediate supervision" of a competent person. This WAH supervisor should be able to intervene immediately if the workers are working unsafely.

The occupier must ensure that all open sides or openings that can cause persons to fall 2 metres or more are covered or guarded by effective guard rails or barriers to prevent falls. If the guard rails or barriers are removed to facilitate work, they must be reinstated as soon as possible, and the workers exposed to falls from height due to the removal of the guard-rails or barriers must be protected by a travel restraint (worker is restrained from reaching the edge or opening) or a fall arrest system (worker's fall can be arrested before he hits the ground or obstacle). The occupier must also ensure that the top guard rail is at least one metre tall and the vertical distance between any two adjacent guard rails is less than 600 mm wide. The guard rail or barrier must also be of good construction, sound material and adequate strength to withstand the impact during the course of fall.

The design of fall arrest and travel restraint systems can be based on Singapore Standard 607:2015 Specification for design of active fall-protection systems (SPRING Singapore, 2015). The fall arrest system must be safe for use, i.e. the force applied on the user body must be less than 6 kN and the user must not hit any obstacle or the ground during the fall. The responsible person overseeing the WAH must appoint a competent person to inspect the anchorage and anchorage line of the travel restraint and fall arrest systems. The inspection by the competent person must be conducted at the start of every shift.

The occupier must ensure that there is always a safe means of access and egress. This includes the staircases that must be effectively barricaded to prevent falls of more than two metres. The openings between the platform and a hoist (aka teagle) must also be securely fenced and provided with secure handholds at the opening or doorway.

Working on roofs, fragile surfaces, and ladders are highly dangerous forms of WAH and they are highlighted in the WAH Regs. For a fixed vertical ladder above nine metres, there must be an intermediate landing place to ensure that the continuous vertical distance does not exceed nine meters.

"Hazardous work at height" refers to work —

(a)   in or on an elevated workplace from which a person could fall a distance of more than 3 metres;

(b)   in the vicinity of an opening through which a person could fall a distance of more than 3 metres;

(c)   in the vicinity of an edge over which a person could fall a distance of more than 3 metres;

(d)   on a surface through which a person could fall a distance of more than 3 metres; or

(e)   in any other place (whether above or below ground) from which a person could fall a distance of more than 3 metres.

All hazardous WAH requires a permit-to-work system to be set up and implemented by the occupier.

The WAH Reg also regulates industrial rope access systems, making requirements like having two independent anchorage lines compulsory and requiring the responsible to engage a professional engineer to design the anchorage and anchorage line of an industrial rope access system.

9.10 Workplace Safety and Health (Workplace Safety and Health Committees) Regulations 2008

The Workplace Safety and Health (WSH Committees) Regulations 2008 ("WSH Committees Regs") stipulates the formation, meetings and functions of a WSH committee, as highlighted in Section 29(1) of the WSH Act. The WSH Committee is required in all factories (high-risk workplaces, as defined in the WSH Act) and the occupier of the factory must set it up. The WSH committee must be chaired by a competent person appointed by the occupier. The secretary is usually the WSH officer and there must be members who are representatives of employees and the management. There must always be more employee representatives in the WSH committee.

The WSH committee must meet at least once a month during office hours to discuss WSH matters and the meetings must be recorded in minutes. The committee must conduct inspections at least once a month. The inspection findings must then be discussed and recorded. Follow-up actions must be implemented. The WSH committee must also conduct an inspection after any accident or dangerous occurrence and the WSH officer must conduct an investigation into the incident and furnish an investigation report to the WSH committee chairman. The post-incident inspection of the WSH committee must be discussed at a meeting and a report recording the WSH issues and recommendations must be produced. The occupier must then evaluate the report and take the necessary actions.

The WSH committee also has other functions, such organising promotion activities to improve WSH, and setting WSH guidelines for the factory. For the WSH committee to be effective, the members must be trained in WSH. The WSH Committee has the following powers:

(a)   to enter, inspect and examine the workplace at any reasonable time;

(b)   to inspect and examine any machinery, equipment, plant, installation or article in the workplace;

(c)   to require the production of workplace records, certificates, notices and documents kept or required to be kept under the Act, including any other relevant document, and to inspect and examine any of them;

(d)   to make such examination and inquiry of the workplace and of any person at work at that workplace as may be necessary to execute his duties;

(e)   to assess the levels of noise, illumination, heat or harmful or hazardous substances in the workplace and the exposure levels of persons at work therein;

(f)   to investigate any accident, dangerous occurrence or occupational disease that occurred within the workplace.

9.11 Workplace Safety and Health (Workplace Safety and Health Officers) Regulations

A WSH Officer must have completed the WSQ Specialist Diploma in WSH and have at least two years of practical experience relevant to WSH. In addition, the WSH Officer must also pass an interview conducted by the Commissioner (Ministry of Manpower). The following workplaces must employ a WSH officer:

(a)   Shipyards in which any ship, tanker and other vessels are constructed, reconstructed, repaired, refitted, finished or broken up.
(b)   Factories used for processing petroleum or petroleum products.
(c)   Factories in which building operations or works of engineering construction of a contract sum of S$10 million or more are carried out.
(d)   Any other factories in which 100 or more persons are employed, except those which are used for manufacturing garments.

The duties of a WSH officer include:

(a)   assisting the occupier of the workplace or other person in charge of the workplace to identify and assess any foreseeable risk arising from the workplace or work processes therein;
(b)   recommending to the occupier of the workplace or other person in charge of the workplace reasonably practicable measures to eliminate any foreseeable risk to any person who is at work in that workplace or may be affected by the occupier's undertaking in the workplace;
(c)   where it is not reasonably practicable to eliminate the risk referred to in sub-paragraph (b), recommending to the occupier of the workplace or other person in charge of the workplace —

    i.   such reasonably practicable measures to minimise the risk; and

    ii. such safe work procedures to control the risk; and

(d)   assisting the occupier of the workplace or other person in charge of the workplace to implement the measure or safe work procedure referred to in sub-paragraph (b) or (c), as the case may be.

As can be observed, the WSH officer is predominantly meant to assist and advice the occupier. To perform their duties, the WSH officer is given the same set of powers as the WSH committee.

References

SPRING Singapore. (2015). *SS 607: 2015 Specification for design of active fall-protection systems*. Singapore: SPRING Singapore.

Tranter, M. (2004). *Occupational Hygiene and Risk Management — Megan Tranter — 9781741143294 — Allen & amp; Unwin — Australia* (2nd ed.). Sidney: Allen & Unwin. Retrieved from https://www.allenandunwin.com/browse/books/general-books/business-management/Occupational-Hygiene-and-Risk-Management-Megan-Tranter-9781741143294

Workplace Safety and Health Council. (2013). *Code of Practice for Working Safely At Heights* (Second Rev). Singapore.

Workplace Safety and Health Council. (2014). Code of Practice on Safe Lifting Operations in the Workplaces. Retrieved March 27, 2018, from https://www.wshc.sg/files/wshc/upload/cms/file/2014/Code_of_Practice_Safe_Lifting_Operations_Revised_2014.pdf

WSH Council. (2017). Asbestos. Retrieved March 19, 2018, from https://www.wshc.sg/wps/portal/!ut/p/a1/jY89D4IwEIZ_iwNr7_gQjVvjIFGMA6jQxYCpB-VMoKRX-vsiqqLfd5XnyvgcMEmB11pUiM6WqM_namX8JD57n0Ah3my-C2kTp7DOKFa0e-NwDpNLA-uf_5ODEUf_lnYN-QscEIfInYAhNS5eO7Ka-1zdymAaX7jmmvy0MO5MKZpVxZa2Pc9EUoJyclVVaQVFn6yCtUaSN5gaKpjgve57EI6ewL

10 Accident Case Studies

10.1 Introduction

In this chapter, two accident case studies are presented to facilitate discussions about key Workplace Safety and Health (WSH) management concepts presented in the earlier chapters. At the same time, relevant WSH legislations will also be highlighted.

10.2 Crane Collapse

10.2.1 Background

This case is taken from *Jurong Primewide Pte Ltd v Moh Seng Cranes Pte Ltd and others [2014] 2 SLR 360*. With reference to Figure 10.1, Jurong Primewide Pte Ltd ("JPW") was the main contractor to build a seven-storey multi-user business park development for Crescendas Bionix Pte Ltd ("Crescendas"). JPW had a crane rental agreement with Hup Hin Transport Co Pte Ltd ("Hup Hin") on a per call basis and a hiring contract with Moh Seng Cranes Pte. Ltd. ("Moh Seng") to hire Moh Seng's mobile cranes whenever required. MA Builders Pte Ltd ("MA") was JPW's subcontractor responsible for structural, architectural and external works.

10.2.2 Accident detail

On 10 June 2010, MA requested JPW to provide a mobile crane to lift some rebars. Thus, JPW contacted Hup Hin to deliver a 50-tonne mobile crane to the worksite the following day. As Hup Hin did not have a 50-tonne mobile crane on 11 June 2010, they contacted Moh Seng for the crane.

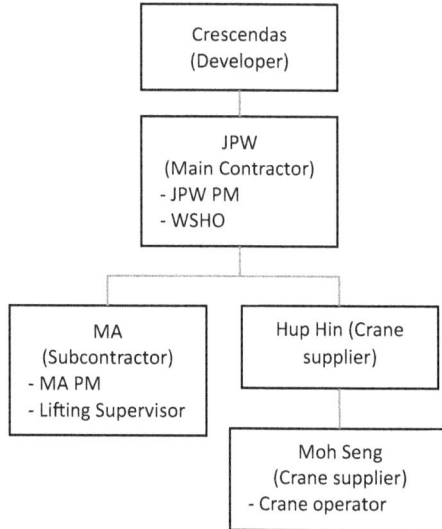

Figure 10.1   Relationship between different parties

On 11 June 2010 the crane operator employed by Moh Seng, arrived at the site with the 50-tonne mobile crane. He reported to the lifting supervisor employed by MA, who instructed the crane operator to park at the washing bay area near the guardhouse, which is in the vicinity of a manhole covered by soil. The crane operator expressed concerns that the designated location might not be able to take the weight of the crane safely. The lifting supervisor assured the crane operator that the location had "hard flooring" and could support the crane's weight. The crane operator was not convinced and he expressed his concerns to JPW's workplace safety and health officer (WSHO). The WSHO discussed with the lifting supervisor and then reassured the crane operator that the ground was able to take the loading imposed by the crane. Thus, the crane operator proceeded to deploy the crane in accordance to the lifting supervisor's instructions.

When the crane operator swung the boom from the left front of the crane towards the left back outrigger, the left back outrigger broke through the cover of a concealed manhole causing the crane to collapse. The manhole was covered up with layers of brown soil and was not visible just prior the accident.

It was established that JPW knew about the manhole by 12 August 2008, but did not take reasonable care to properly mark and cordon off the manhole. They also did not direct the subcontractors to take the appropriate action. JPW indicated that the manhole was usually visible and so "obvious to any person or passer-by" that there would not have been "the need of [sic.] any barricade, warning sign or warning". They also blamed MA for allowing soil to accumulate over the manhole due to their excavation work. However, the Judges found that the sheer knowledge of the presence of the manhole would have made it foreseeable that the accident could happen on a busy site with heavy equipment and machinery. Thus, JPW's negligence was *inexcusable*.

Nevertheless, MA, as subcontractor, was also expected to ensure that the lifting operation was conducted safely. This is especially the case because MA was the employer of the lifting supervisor who was instructing the lifting work. Furthermore, MA was fully informed of the location of the manhole because they conducted investigation and piping works in the manhole prior to the accident. MA's project manager (PM) had also warned the lifting supervisor about the manhole, and the lifting supervisor acknowledged that he was aware of the presence of the manhole. Despite this knowledge, MA's lifting supervisor commenced lifting operations without a valid permit-to-work (PTW).

10.2.3 ECT analysis

Figure 10.2 shows the Event Causation Technique (ECT) diagram for the accident. As seen in Fig. 10.2, the breakdown event is in box (b), "Left back outrigger placed over concealed manhole". The direct cause of the accident was the decision made by the lifting supervisor and JPW WSHO to locate the crane near the washing bay. This decision was probably based on their assumption that the manhole cover could take the load from the outrigger and/or the outrigger was not placed directly over the concealed manhole. The lifting supervisor and WSHO could have been improperly motivated, but there was a lack of information on the reason why the two safety personnel made the decision despite concerns from

Date of Accident: 11-Jun-10 (Friday)

Work Context: 50-tonne mobile crane used to lift rebar at construction site

Incident Sequence

Direct Causes

Control Failures

Underlying Factors

(a) Moh Seng's crane being set up for lifting of rebar; crane boom swung from left front of crane to left back outrigger

(b) Left back outrigger placed over concealed manhole

(c) Left back outrigger broke through manhole cover

(d) Crane collapsed

(e) Mobile crane damaged; manhole damaged; no injury (?)

(f) MA lifting supervisor and JPW WSHO assumed that manhole cover can take load from outrigger and/or outrigger not over manhole; Improper motivation (?)

(g) Manhole concealed by soil from nearby excavation by MA

(h) Permit-to-work for lifting not implemented

(i) Failure to mark and cordon off manhole

(i) *Lifting supervisor and WSHO ignored crane operator's concern, when the manhole is known to be in the vicinity; shows unsafe attitude among safety personnel. This implies a poor state of safety culture and leadership.*

(k) *PTW system not implemented implies that operational controls of the WSH management system are not adequately implemented.*

Legend

Incident Event

Control Failures

Direct cause

Underlying factors

? - inadequate/uncertain/conflicting evidence

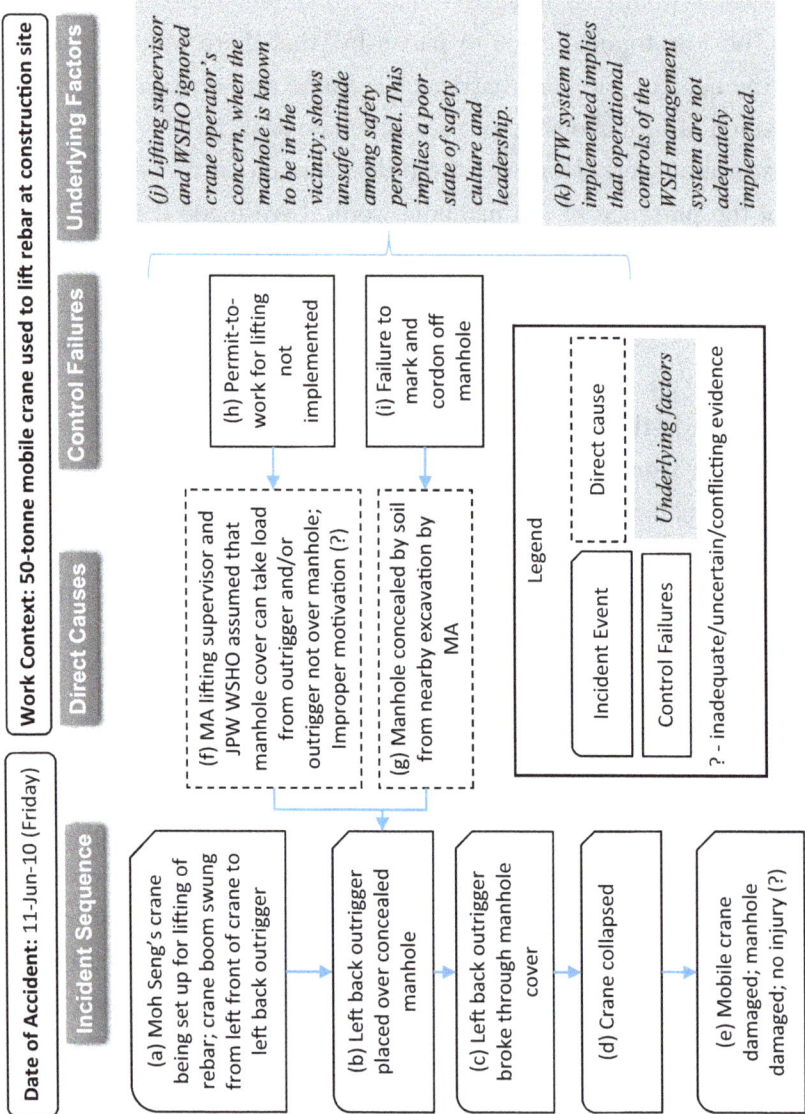

Figure 10.2   Event causation technique (ECT) diagram

the crane operator. The concealment of the manhole was also a direct cause leading to the outrigger being placed over the concealed manhole.

One control that could have been implemented to prevent the breakdown event is a PTW system to ensure that legal requirements such as checking on ground conditions were conducted prior to lifting. Reg 12 of the Workplace Safety and Health (Construction) Regulations 2007 (No S 663/2007) required a PTW to be first issued by the PM of the worksite before high-risk construction work could be conducted. At the time of the accident, Reg 20(3)(*c*) of the Factories (Operations of Cranes) Regulations 1998 ("the Regulations") was still in force and it stated that a lifting supervisor has to "ensure that the ground conditions are safe for any lifting operation to be performed by any mobile crane". This was also highlighted in the Singapore Standard SS 536 2008 Code of Practice for the safe use of mobile cranes, which specifically identified manholes as one of the possible "underground hazards" that must be checked for.

Another possible control that could have prevented the outrigger from being placed near or over the manhole is to clearly mark and cordon off the manhole area. This could also potentially prevent the manhole from becoming concealed by the soil during excavation.

Through the actions of the lifting supervisor and WSHO, who ignored the concerns of the crane operator about the ground condition, it is inferred that the safety culture and leadership of the site was probably lacking prior to the accident. In addition, with the failure to implement the permit-to-work system for high-risk activities like lifting, it can be inferred that the level of implementation of operations control element of the management system was probably inadequate.

10.2.4 Discussion

In their judgement, the Appeal Judges highlighted the intention of the Workplace Safety and Health Act (WSHA), which is to create a system of accountability by defining the duties of different stakeholders. The then Minister of Manpower Dr Ng Eng Hen during the second reading of the Workplace Safety and Health Bill 2005 (Bill 36 of 2005)

(Singapore Parliamentary Debates, Official Report (17 January 2006) vol 80) at cols 2208–2210 explained:

> "Third, this Bill will better *define persons who are accountable, their responsibilities and institute penalties* which reflect the true economic and social cost of risks and accidents. Penalties should be sufficient to deter risk-taking behaviour and ensure that companies are *proactive in preventing incidents.* Appropriately, companies and persons that show *poor safety management* should be penalised even if no accident has occurred.
>
> This Bill will put into place a new and more effective framework to reduce accidents at the workplace — to bring about a quantum improvement in OSH standards and to achieve our intermediate goal of halving the present occupational fatality rate by 2015."

Accordingly, contractors and subcontractors, i.e. JPW and MA respectively, are two of the entities which the WSHA seeks to "increase direct liability on for workplace safety". Main contractors and subcontractors have direct "operational control" of workplaces and they have tremendous responsibilities to ensure a safe working environment at construction sites. Even though the main contractor, as the Occupier of the worksite, has a duty under Section 11 of the Workplace Safety and Health Act (Chapter 354A) to ensure, so far as is reasonably practicable, that the workplace, all means of access to or egress, and any machinery, equipment, plant, article or substance are safe, principals (MA in this case) and employers have an important role because they direct the manner of work. Dr Ng Eng Hen, in the second reading of the Workplace Safety and Health Bill 2005, also explained how the WSHA sought to impose duties on principals engaging contractors for specialised tasks (at col 2209):

> "Traditionally, a principal who engages a contractor would be engaging the specialist services of the contractor, and would not be directing the contractor on how to do the work. However, today the situation is different. Principals often engage "contractors" and third-party labour

not for their specialist expertise, but *precisely so that they can avoid entering into a direct employment relationship, for organisational or other reasons.* In such situations, the principal, in terms of supervision, takes on the role of an employer. *The Bill thus places on him responsibility for the worker's safety and health as if he were the employer.* If this were not the case, then the *duties under the Act could be simply circumvented by a careful crafting of the legal relationship.*"

Thus, JPW, as the main contractor, had to ensure that "hazards such as manholes were identified properly through site surveys, and then to undertake ground improvements to ensure that these underground hazards did not continue to remain hazardous". Even though MA was "not contributorily negligent in causing some soil run-off to cover the manhole due to its nearby excavation works", it is still responsible for the accident because the "lifting operation that caused the accident was within MA's scope of work" (the lifting supervisor, riggers and signalmen were employed by MA), "MA knew about the manhole as well", and "MA was contractually responsible and deemed fully informed of the conditions at the worksite".

The lifting supervisor is an important safety personnel during a lifting operation, but in this case the lifting supervisor failed to perform his duties specified in Reg 20(3)(c) of the Factories (Operations of Cranes) Regulations 1998, which states that a lifting supervisor has to "ensure that the ground conditions are safe for any lifting operation to be performed by any mobile crane". Furthermore, the Singapore Standard SS 536 2008 Code of Practice for the safe use of mobile cranes, required the lifting supervisor to:

(a) *co-ordinate and be present to supervise all lifting activities and ensure that the lifting operation is carried out safely;*

(b) ensure that only registered crane operators, appointed riggers and appointed signalmen participate in any lifting operation involving the use of a mobile crane;

(c) *ensure that the ground conditions are safe for any lifting operation to be performed by any mobile crane;*

(d) ensure that there is a set of safe lifting procedures for any lifting operation of a mobile crane;

(e) brief all crane operators, riggers and signalmen on the safe lifting procedures referred to in (d);

(f) ***take measures to rectify the unsatisfactory or unsafe conditions that are reported by any crane operator or rigger.***

The Judges were especially critical on the lifting supervisor because though he was aware that there was a concealed manhole, he failed to ensure that the hazard was effectively managed. Worse still, the lifting supervisor failed to take any measures to address the crane operator's concerns about the safety of the ground conditions. Thus, the Judges "had no hesitation in finding that MA was liable in negligence for the damage caused during the accident".

Finally, the Judges made the following decision: "both JPW and MA were liable in negligence to Moh Seng, and that MA did breach the subcontracts that it had entered into with JPW, we felt that an apportionment of 60% to JPW and 40% to MA would be just in all the circumstances. As the main contractor, as well as the occupier under the WSHA, JPW had to bear the bulk of the responsibility for failing to identify an underground hazard like the manhole and for not taking measures to ensure that it ceased to be an unknown danger. However, we were also cognisant of the fact that MA was largely in charge of the works that were in the area, and specifically, the operation of lifting steel rebars that [the crane operator's] crane was engaged in. Also, the Lifting Supervisor was under MA's employment and had specific duties under the WSH Regime which he failed to fulfil."

10.3 Falling From Loading Platform During Lifting

10.3.1 Background

This case is taken from *Public Prosecutor v GS Engineering & Construction Corp [2016] SGHC 276*. With reference to Figure 10.3, GS

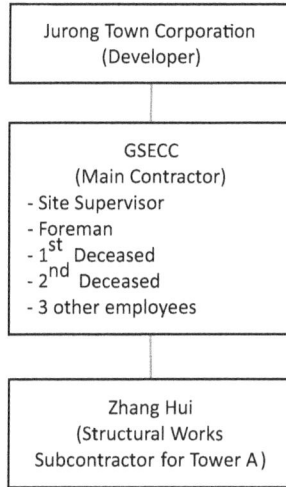

Figure 10.3   Parties involved

Engineering & Construction Corp (GSECC) was the main contractor to construct two towers (Tower A and Tower B) for Jurong Town Corporation. The two towers were to be 11 and 18 storeys high respectively. GSECC subcontracted the structural works (including formwork) for Tower A to Zhang Hui Construction Pte Ltd ("Zhang Hui"). The project started on 23 November 2011 and was meant to be completed by 23 March 2014.

10.3.2 Accident detail

On 22 January 2014, GSECC employees were initially planning to shift the loading platform from the tenth floor of Tower B to the eighth floor of Tower A. Loading platforms are typically positioned on the edge of buildings to facilitate the lifting of materials and items to and from different levels of the building. When employees from Zhang Hui requested the GSECC site supervisor to move the air compressor using the loading platform, the site supervisor instructed the GSECC foreman to assist Zhang Hui on this task. The site supervisor instructed the foreman "not to install the loading platform at the seventh storey of Tower A, but to simply suspend it by a tower crane."

At about 11.50am, the foreman with five workers, including the two deceased, commenced the task of shifting the loading platform from Tower B to the seventh storey of Tower A. As Zhang Hui's workers were at lunch, the GSECC foreman proceeded with his workers. The air compressor was mounted on a steel frame with four wheels, but the rear wheels were smaller and could not roll onto the platform. The workers then used a galvanised pipe to pivot the air compressor onto the loading platform. After several attempts, the air compressor was pushed onto the platform, but it rolled towards the two deceased who were on the platform, causing the platform to tilt. The two deceased were not able to evade the air compressor in time and fell together with it.

The air compressor landed on another loading platform two levels below. The two deceased fell to the ground level and were pronounced dead by the paramedics attending to the accident.

10.3.3 ECT analysis

The incident sequence is summarised in boxes (a) to (e) of Figure 10.4. The breakdown event is described in box (b) "Air compressor rolled towards 2 deceased" and the contact event is described in box (e) "2 deceased hit ground".

The breakdown event happened because prior to it, the workers were pushing the air compressor on wheels onto the platform and the loading platform was suspended by the tower crane using four lifting chain slings. Effective training on the safe use of the loading platform (box (i)), presence of a competent lifting supervisor (box (j)), implementation and briefing of the permit-to-work (PTW), risk assessment (RA) and safe work procedure (SWP) for lifting (box (k)), and conducting RA and SWP for the loading platform (box (l)) should have resulted in the removal of the unsafe work method and hence prevented the breakdown event. In addition, the lack of provision of fall protection equipment (fall arrest anchor, lifeline and harness) to the workers on the day of the accident allowed the contact event (box (d)) to happen.

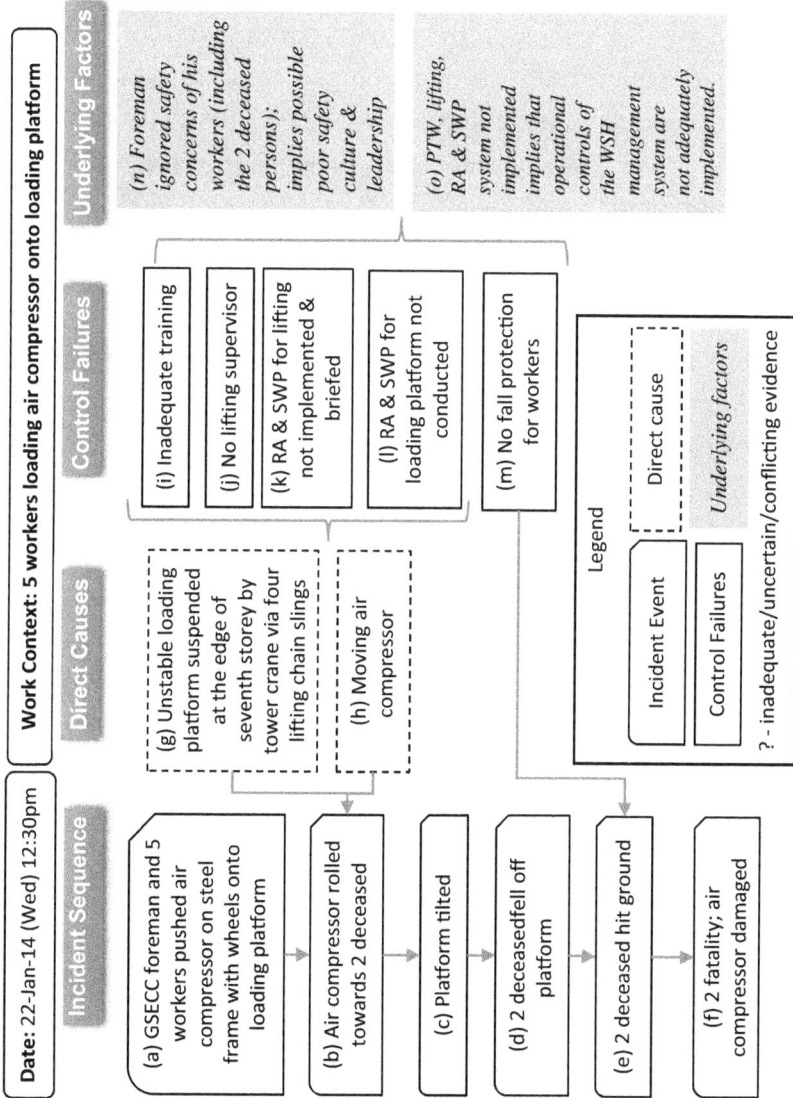

Figure 10.4 ECT diagram for GSECC v PP

Date: 22-Jan-14 (Wed) 12:30pm

Work Context: 5 workers loading air compressor onto loading platform

Underlying Factors

(n) Foreman ignored safety concerns of his workers (including the 2 deceased persons); implies possible poor safety culture & leadership

(o) PTW, lifting, RA & SWP system not implemented implies that operational controls of the WSH management system are not adequately implemented.

Control Failures

(i) Inadequate training

(j) No lifting supervisor

(k) RA & SWP for lifting not implemented & briefed

(l) RA & SWP for loading platform not conducted

(m) No fall protection for workers

Direct Causes

(g) Unstable loading platform suspended at the edge of seventh storey by tower crane via four lifting chain slings

(h) Moving air compressor

Incident Sequence

(a) GSECC foreman and 5 workers pushed air compressor on steel frame with wheels onto loading platform

(b) Air compressor rolled towards 2 deceased

(c) Platform tilted

(d) 2 deceased fell off platform

(e) 2 deceased hit ground

(f) 2 fatality; air compressor damaged

Legend

Incident Event

Control Failures

Direct cause

Underlying factors

? - inadequate/uncertain/conflicting evidence

The unsafe conditions were identified by the workers, and they communicated their concerns to the foreman prior to the accident. However, the foreman failed to address their concerns and instructed them to proceed with the task (box (n)). There is a lack of information on the state of safety culture of the project, but this piece of evidence indicates that frontline leaders (supervisors and foremen) may not place due emphasis on safety. Similarly, with several operational controls like PTW, RA and SWP not adequately implemented, it is a sign that the management system is ineffective.

10.3.4 Discussion

This case is a landmark case because it was the first time a WSHA case was presented to the High Court. Furthermore, the Prosecution disputed the District Judge's sentencing and proposed a new sentencing framework. One of the key rationales for the Prosecution's proposal was that the sentencing for previous cases were too low, with the majority falling below 30% of the maximum sentence of S$500,000 prescribed in the WSHA. The relatively low sentencing was contrary to the intent of the WSHA. The then-Minister for Manpower, Dr Ng Eng Hen, at the second reading of the Workplace Safety and Health Bill (Bill 36 of 2005) ("the Bill"), described the legislative intent of the WSHA:

> "Three fundamental reforms in this Bill will improve safety at the workplace. First, the Bill will strengthen proactive measures. Instead of reacting to accidents after they have occurred, which is often too little too late, we should reduce risks to prevent accidents. To achieve this, all employers will be required to conduct comprehensive risk assessments for all work processes and provide detailed plans to minimise or eliminate risks.
>
> Second, industry must take ownership of occupational safety and health standards and outcomes to effect a cultural change of respect for life and livelihoods at the workplace. …

Third, this Bill will better define persons who are accountable, their responsibilities and *institute penalties which reflect the true economic and social cost of risks and accidents. Penalties should be sufficient to deter risk-taking behaviour and ensure that companies are proactive in preventing accidents.* Appropriately, companies and persons that show poor safety management should be penalised even if no accident has occurred."

Furthermore, Dr Ng explained the reason for the need for higher penalties for poor safety management and performance:

"Even as we work with industry to build up their capabilities to improve safety and health at their workplaces, *we need to ensure that the penalties for non-compliance are sufficiently high to effect a cultural change on the ground. Penalties should be set at a level that reflects the true cost of poor safety management, including the cost of disruptions and inconvenience to members of the public which workplace accidents will cause.* The collapse of the Nicoll Highway not only resulted in the loss of four lives, but also caused millions of dollars in property damage and led to countless lost working hours and great inconvenience to the public. The maximum penalty of $200,000 under the present Factories Act is therefore inadequate."

The High Court Judge largely adopted the proposed sentencing framework, which is presented in Table 10.1.

Table 10.1 Sentencing framework determined by High Court

		Culpability		
		High	**Medium**	**Low**
Potential for harm	High	$300,000 to $500,000	$150,000 to $300,000	$100,000 to $150,000
	Medium	$100,000 to $150,000	$80,000 to $100,000	$60,000 to $80,000
	Low	$40,000 to $60,000	$20,000 to $40,000	Up to $20,000

The Judge also provided a detailed description of the process for sentencing by a court using Table 10.1. With reference to Table 10.1, a court will determine the appropriate starting point for the sentence by considering two principal factors: (i) the culpability of the offender; and (ii) the harm that could potentially have resulted, as described below:

1. Determine the culpability of the offender based on the following non-exhaustive factors:

   a. the number of breaches or failures in the case;
   b. the nature of the breaches;
   c. the seriousness of the breaches — whether they were a minor departure from the established procedure or whether they were a complete disregard of the procedures;
   d. whether the breaches were systemic or whether they were part of an isolated incident; and
   e. whether the breaches were intentional, rash or negligent.

2. The potential harm may be assessed by considering, among other things, (i) the seriousness of the harm risked; and (ii) the likelihood of that harm arising.

3. After deriving the starting point for sentencing, the court should calibrate the sentence by taking into account the aggravating factors and mitigating factors of the case.

4. Aggravating factors include the following: (i) serious actual harm (including death) resulted; (ii) the breach was a significant cause of the harm that resulted — in this regard, a significant cause need not be the sole or principal cause of the harm, and need only be a cause that has more than minimally, negligibly or trivially contributed to the outcome; (iii) the offender had cut costs at the expense of the safety of the workers; (iv) there was a deliberate concealment of the illegal nature of the activity; (v) there was a breach of a court order; (vi) there was an obstruction of justice; (vii) the offender has a poor record with respect to workplace health and safety; (viii) there was

falsification of documentation or licences; and (ix) there was a deliberate failure to obtain or comply with relevant licences in order to avoid scrutiny by the authorities.

5. Mitigating factors may include the following: (i) the offender has voluntarily taken steps to remedy the problem; (ii) the offender provided a high level of cooperation with the authorities for the investigations, beyond that which is normally expected; (iii) there is self-reporting, cooperation and acceptance of responsibility; (iv) there is a timely plea of guilt; (v) the offender has a good health and safety record; and (vi) the offender has effective health and safety procedures in place.

For the case of *GSECC v PP*, the Judge opined that GSECC's (Respondent) culpability falls into the Medium to High category because the occupier is overall in-charge of the worksite. Concurrently, GSECC was also the employer of the two deceased and the other workers involved in the task that resulted in the accident. Furthermore, the occupier failed to perform numerous control measures that are expected of their own lifting operations and it also failed to ensure that its workers were adequately trained and had adequate fall protection. Thus, even if Zhang Hui had not asked GSECC to shift the air compressor, the occupier would have committed numerous breaches and failures. Moreover, the occupier oversees the PTW system and would have to evaluate and approve Zhang Hui's application to conduct the lifting operation.

During the trial, the Respondent attempted to shift the responsibility to its workers, but the Judge noted the then Minister of Manpower, Dr Ng Eng Hen's, observations at the second reading of the Workplace Safety and Health Bill (*Singapore Parliamentary Debates, Official Report* (17 January 2006) vol 80 at col 2205):

"The reality is that on a day-to-day basis, safety may be the last thing on the minds of management and workers on the ground. There are deadlines to meet, monotony, apathy or lethargy to overcome, a lack of professionalism and training, unclear lines or no lines of accountability, and poor management…"

Even though workers have duties under the WSHA to cooperate with their employers, the intent of the WSHA is for the employer, occupier or other responsible persons to "ensure that its workers are trained and are mindful of their safety at the workplace, and that proper systems are in place to ensure that steps are taken to minimise risks."

In terms of potential harm, the Judge selected the "high" category because the high-risk work was conducted without a safe system of work, and fall arrest equipment were not provided. Furthermore, many of the workers were not trained for the task. Thus, all the workers involved could have been killed. In addition, the landing platform could have failed and landed on other workers.

Therefore, based on the level of culpability and the potential for harm, the starting sentence was a fine of S$300,000.

The Judge then considered the aggravating and mitigating factors to calibrate the sentence. Aggravating factors included two lives being lost as a result of the breaches. Mitigating factors included the fact that the Respondent had pleaded guilty, cooperated with the authorities, had a good safety record, won past awards from the Land Transport Authority and MOM (two WSH awards for the accident site), and had been proactive in investigating the accident and implementing remedial actions. Finally, after weighing all the relevant factors, the Judge arrived at an appropriate sentence of S$250,000.

10.4 General Discussion

The two accident cases presented important WSH management lessons. These cases highlight the importance of front line leadership by supervisors, foremen and site personnel. These supervisory personnel make many decisions on the ground that can save or kill workers, and their decisions will be based on their perception of what is important and what they believe is the right way to conduct their work. Applying the concepts discussed in earlier chapters, to ensure that supervisory personnel adhere to safety procedures, it is of utmost importance that managers impress upon them the importance of WSH. To do so managers must have dedicated WSH communication sessions with workers and supervisors.

At the same time managers must "give meaning" to the administrative controls such as permit-to-work systems, risk assessments and safe work procedures, which, as can be observed in the two cases discussed in this chapter, can be easily defeated and violated. It is a common problem on worksites that paperwork is just paperwork, i.e. the controls identified are not implemented and communicated to the workers. It is only when managers review and discuss the documents with site personnel then they will see the relevance of these documents. If the documents are used and referred to, the quality of risk assessment and control measures will be improved.

At the same time, supervisory personnel must be trained on soft skills like communication and team leadership so that they can effectively tap into the knowledge and wisdom of the workers under their charge to effectively prevent accidents. As can be seen from the two cases, workers on the ground can foresee possible accidents, and failure to listen to workers' concerns can result in accidents. Supervisory personnel should address all safety concerns through the risk assessment process. When a worker identifies a hazard, the work team should stop their work and discuss the potential consequences and their likelihood so that suitable measures can be implemented to prevent accidents and ill health. Such onsite and quick risk assessment forces the supervisor or foreman to take a systematic and detailed look at the task instead of addressing the hazards intuitively, during which he may be prone to cognitive biasness such as over-confidence.

The range of possible hazards and accidents is very wide. Thus, when evaluating and learning from accident cases, it is important to consider the underlying factors. However, most investigations do not have the resources to probe the management system, culture and leadership issues in sufficient depth. As can be seen from the two case studies, only subjective inferences can be made. Nevertheless, the violation of WSH rules and procedures and the failure to listen and communicate do point to poor safety culture. Leadership and management system are common issues in many organisations and they are still the key to eradicating accidents and ill health in workplaces.

Index